ESSENTIALS OF
ABSTRACT ALGEBRA

CONTEMPORARY UNDERGRADUATE MATHEMATICS SERIES
Robert J. Wisner, Editor

ESSENTIALS OF ABSTRACT ALGEBRA

CHARLES M. BUNDRICK

University of West Florida

JOHN J. LEESON

University of North Florida

BROOKS/COLE PUBLISHING COMPANY

Monterey, California

A Division of Wadsworth Publishing Company, Inc.

Belmont, California

This book was designed by Linda Marcetti, and
its production was supervised by Phyllis London.
The technical art was drawn by Reese Thornton.
The book was printed and bound by Kingsport
Press, Kingsport, Tennessee.

QA
266
B79

ISBN: 0-8185-0043-3
L.C. Catalog Card No: 75-187499
Printed in the United States of America

1 2 3 4 5 6 7 8 9 10—76 75 74 73 72

To Pat and Teresita

PREFACE

Essentials of Abstract Algebra is designed for a first course in abstract algebra. The text was developed from material presented by the authors to sophomore- and junior-level mathematics majors, some of whom were preparing for careers as high school teachers.

Using the basic algebraic properties of the integers, the rational numbers, and the real numbers as a source of illustrative examples, we proceed to abstract the basic notions comprising modern algebra. The unifying concepts of modern algebra provide a natural setting for demonstrating the power and substance of abstract mathematical thought. And, through examples, reference to previous mathematical experiences shows how abstractions arise out of concrete situations. Therefore in order to understand the examples, the student should have had some successful experience with college-level mathematics. Other than this necessary experience, there are no formal prerequisites for use of the book.

Chapter 1 is designed to reinforce and equalize the readers' mathematical backgrounds. Even for well-prepared students, we suggest a cursory reading of at least the last three sections of Chapter 1, since some notational conventions are introduced there. The entire text includes enough material for about six or seven quarter hours of work—five or six if Chapter 1 is skimmed or omitted.

As with all textbooks in mathematics, the exercises are an extremely important part of the learning process. Many easy exercises are included so that the student may become familiar with new notations and concepts. More challenging exercises are also included to allow the student to become involved with the ideas that we have presented. Answers to selected problems can be found at the end of the book. The symbol † is used to identify these exercises.

To help the student use the text as a reference, we have identified definitions, theorems, and tables by three numbers. The first number refers to the chapter, the second to the section, and the third to the particular theorem, definition, or table within the section. Occasionally a reference is made to a theorem divided into two or more parts. Such references use four numbers with the last one indicating the appropriate part of the theorem. The symbol ▲ indicates the end of a proof.

We have stressed the use of mappings as an investigatory aid in the study of familiar and abstract algebraic systems. In addition, we use an outline-type division of long, detailed proofs. This is done so that, hopefully, the overall picture will not be lost in the necessary details.

In the study of group theory, results are presented in a multiplicative environment. In ring theory, these same results are applied in an additive environment. The last section of Chapter 4 helps bridge this gap by restating the pertinent group-theoretical results in additive notation. Throughout, we give concrete examples of new concepts to help cement the relationship between the familiar and the abstract.

Our thanks go to many who have helped with the preparation of this book. Special thanks are due Sharon Zeier, who typed most of the manuscript, and Jack N. Thornton and Robert J. Wisner of Brooks Cole Publishing Company. Thanks also to William E. Briggs, University of Colorado, James Jackson, College of San Mateo, Roy Mazzagatti, Miami Dade Junior College, A. Richard Mitchell and Roger Mitchell, University of Texas at Arlington, and David J. Rodabaugh, University of Missouri, for their helpful reviews.

CONTENTS

1 THE LANGUAGE OF MATHEMATICS   1

ESSENTIALS OF ABSTRACT ALGEBRA

THE LANGUAGE OF MATHEMATICS

Two of the main difficulties in a first encounter with abstract mathematics are (1) mathematical discourse usually involves a tremendous amount of notation and (2) there is a necessity to combine abstract ideas in some systematic way to "prove" things. This chapter is devoted to introducing some of the necessary notation and providing some insight into the second area of difficulty. The initial goal is to study symbolic logic for its usefulness as a tool in the construction of mathematical proofs. For this reason, the development herein is informal.

1.1 PROPOSITIONS

Statements used in mathematics are of a rather precise nature and are required to have the property that they are either true or false. There is no "judging" permitted in making this true or false assignment to a given statement. We restrict our attention solely to statements with a unique truth value and call such statements *propositions*.

Consider the following statements.

1. John Jones is 5 years old this year.
2. That rose is red.
3. It is warm outside today.
4. John Jones lives in Florida.
5. Water is H_2O to a chemist.
6. $3 + 2 = 5$

Assuming that the name John Jones identifies the same person for each of us, we can assign exactly one truth value to Statement 1 simply by comparing the date on a current calendar with his birth certificate. Statement 4 can be checked by looking at his address and is certainly either true or false. High school chemistry establishes the truth value for Statement 5, while all would agree that Statement 6 is true. Each of the remaining statements (2 and 3), however, requires a judgment decision that could go one way or the other, depending on one's own point of view. Statement 2 can be assigned different truth values for a given rose according to whether one considers it to be pink or light red. Similarly, judgment is required in considering Statement 3. A warm day to an Alaskan might be considered downright cold by a Floridian. Statements 1, 4, 5, and 6 are propositions, whereas Statements 2 and 3 are not.

There are many ways of combining simple propositions to form new, more complex statements called *compound propositions*. We consider only the types of statements commonly used in mathematics. The connectives that we shall study, along with the usual notation, are

$\backslash$ represents the *negative* "not"
$\wedge$ represents the *conjunction* "and"
$\vee$ represents the *disjunction* "or"
$\leftrightarrow$ represents the *equivalence* "if and only if"
$\rightarrow$ represents the *conditional* "if ..., then ...".

Let p denote the proposition "2 is a prime integer" and q denote the proposition "2 is an even integer." The compound propositions

\\p, $p \wedge q$, and $p \vee q$ become, respectively, "2 is not a prime integer," "2 is a prime integer and 2 is an even integer," and "2 is a prime integer or 2 is an even integer." The conditional $p \to q$ (also called the *implication*) becomes "if 2 is a prime integer, then 2 is an even integer" (also "2 is a prime integer implies 2 is an even integer"). The conditional in symbolic form is usually read as p implies q.

In compound propositions such as $p \wedge q$, $p \to q$, and $p \leftrightarrow q$, p and q are called *component propositions*. The truth value of a compound proposition is determined by the truth values of its component propositions. Consider $p \wedge q$. Proposition p can be either true or false and so can q. Thus four possible pairs of truth values must be considered, and in each case we want a truth value for $p \wedge q$. The truth table in 1.1.1 formally defines the truth values of $p \wedge q$ and of the other familiar compound propositions that use the basic connectives.

1.1.1 TABLE

p	q	\\p	$p \wedge q$	$p \vee q$	$p \leftrightarrow q$	$p \to q$
T	T	F	T	T	T	T
T	F	F	F	T	F	F
F	T	T	F	T	F	T
F	F	T	F	F	T	T

Most of the definitions in 1.1.1 seem "intuitively sound." It is reasonable to assign a true truth value to $p \wedge q$ only when p and q are both true because the conjunction "and" is used in this way in ordinary conversation. Similarly, if we agree to use the disjunction "or" in its inclusive sense of "one or the other or both" (at least one), then the definition for $p \vee q$ seems soundly based. If we interpret equivalence to mean "essentially the same as," then assigning a true truth value to $p \leftrightarrow q$ only when p and q have the same truth value is again compatible with our intuition. This interpretation seems sound, since equivalence and "sameness" are somewhat synonymous. The negative "not" is self-explanatory.

The conditional $p \to q$ seems to stand out as arbitrary and unintuitive, at least in part. Good judgment warrants a true truth value for $p \to q$ when both p and q are true and a false truth value when p is true and q is false. However, why do we define $p \to q$ to be true when p is false? Have we made a completely arbitrary definition, or is there some underlying reason? Our intuition fails to help at this point, since, in ordinary conversation, we seldom use the conditional $p \to q$ unless p is true. Some truth value must be assigned to $p \to q$ in each of these cases; otherwise $p \to q$ would fail to be a proposition. All the possible combinations for $p \to q$ (assuming the first two rows are soundly based) are listed in 1.1.2.

1.1.2 TABLE

		A	B	C	D
p	q	$p \to q$	$p \to q$	$p \to q$	$p \to q$
T	T	T	T	T	T
T	F	F	F	F	F
F	T	T	T	F	F
F	F	T	F	T	F

If column B had been chosen as the definition for $p \to q$, then q and $p \to q$ would always have the same truth values. This situation would not be acceptable, since these two concepts do not have the "intuitive sameness" that equivalence requires. Similarly, a selection of either column C or column D to define $p \to q$ would lead to a duplication of one of the other basic compound propositions and would thus be intuitively unacceptable. Virtually by elimination we are led to column A as the desirable choice for the definition of $p \to q$.

Determining the truth table of a more complicated compound proposition is just a matter of building from simple propositions by applying the basic compound proposition definitions as we build.

EXAMPLES

1. Construct a truth table for $p \vee \backslash q$.

p	q	$\backslash q$	$p \vee \backslash q$
T	T	F	T
T	F	T	T
F	T	F	F
F	F	T	T

2. Find the truth values for $(\backslash p \vee q) \to p$.

p	q	$\backslash p$	$\backslash p \vee q$	$(\backslash p \vee q) \to p$
T	T	F	T	T
T	F	F	F	T
F	T	T	T	F
F	F	T	T	F

3. Construct a truth table for $(p \to q) \leftrightarrow (\backslash p \vee q)$.

p	q	$p \to q$	$\backslash p$	$\backslash p \vee q$	$(p \to q) \leftrightarrow (\backslash p \vee q)$
T	T	T	F	T	T
T	F	F	F	F	T
F	T	T	T	T	T
F	F	T	T	T	T

In Example 3, the truth values of $(p \to q) \leftrightarrow (\backslash p \vee q)$ are all true. Thus the truth value assigned to this compound proposition is independent of the truth values assigned to the initial component propositions. Propositions with this property are very important in mathematics.

1.1.3 DEFINITION. A compound proposition that is true regardless of the truth values of its initial components is called a *tautology*.

An extremely useful consequence of our definition of $p \to q$ is that $[(p \to q) \wedge (q \to p)] \leftrightarrow (p \leftrightarrow q)$ is a tautology (see Exercise 5). Therefore, the equivalence can be considered as a two-way implication, and indeed the symbol $\leftrightarrow$ is appropriate.

EXERCISES

1. Find the initial components in each of the following compound statements.
   (a) The sun is shining and it is raining.
   (b) It will rain tomorrow if and only if we go to the beach.
   (c) The polygon is either a triangle or a square.

   (d) If Jimmy has a headache and his eyes are red, then he stayed up late
       last night.

   (e) Jimmy did not stay up late last night.

†2. Write the following in symbolic form if p is "n is an odd integer" and q is
    "n is a prime integer."

   (a) If n is not a prime integer, then n is an odd integer.

   (b) n is an odd integer and a prime integer.

   (c) n is an odd integer if and only if n is a prime integer.

3. Let p be "John is lucky" and q be "Tom is handsome." Give a verbal
   translation for
   (a) $p \wedge q$,   (b) $p \wedge \backslash q$,   (c) $\backslash(p \wedge q)$,   (d) $\backslash(p \vee q)$,   (e) $p \vee \backslash q$.

4. Construct truth tables for the symbolic statements in Exercise 3. Can you
   find alternate symbolic statements for parts (c) and (d)?

†5. Construct a truth table to show that $[(p \rightarrow q) \wedge (q \rightarrow p)] \leftrightarrow (p \leftrightarrow q)$ is a
    tautology.

†6. We defined $\vee$ in the inclusive sense. Give a compound statement that
    symbolically states "p or q but not both," using only $\backslash$, $\vee$, and $\wedge$.

1.2   DIRECT PROOF

Most of the results (theorems) in any mathematics text are in
the form of if..., then... statements (or if and only if statements that
separate into two if..., then... statements). Therefore when we start
"constructing proofs," we want to be quite familiar with the connective $\rightarrow$.

The conditional differs from the conjunction, the disjunction,
and the equivalence in the sense that it lacks symmetry. The truth tables
for $p \wedge q$ and $q \wedge p$ are identical (as are those for $p \vee q$ and $q \vee p$,
and for $p \leftrightarrow q$ and $q \leftrightarrow p$). Such is not the case for $p \rightarrow q$ and $q \rightarrow p$.
The conditional $q \rightarrow p$ is called the *converse* of $p \rightarrow q$ and has truth
values as listed in 1.2.1.

1.2.1 TABLE

p	q	$p \rightarrow q$	$q \rightarrow p$
T	T	T	T
T	F	F	T
F	T	T	F
F	F	T	T

Errors in reasoning often occur because of a confusion between the conditional and its converse.

The proof of a theorem stated in the form $p \to q$ may proceed in many different ways. A direct proof follows a path of sound deductive reasoning and leads from the *hypothesis* p to the *conclusion* q. This sounds great; but how do we construct a path consisting of logically valid steps? One important way is to model our argument after a tautology. Clearly this "method of proof" is indisputable, and it eliminates errors that might otherwise be caused by variations in the truth values of the hypotheses with which we begin.

The basic tautology used in the construction of a direct proof is $[(p \to q) \land p] \to q$. In order to stress the importance of this result, we state it as our first theorem.

1.2.2 THEOREM. The compound proposition $[(p \to q) \land p] \to q$ is a tautology.

Proof: The construction of 1.2.3 suffices as a proof of 1.2.2. ▲

1.2.3 TABLE

p	q	$p \to q$	$(p \to q) \land p$	$[(p \to q) \land p] \to q$
T	T	T	T	T
T	F	F	F	T
F	T	T	F	T
F	F	T	F	T

This tautology guarantees that when the hypothesis p is true, we only need to gain the acceptance of $p \to q$ to conclude that q is true. (*Caution:* The tautology does not make the blanket statement that q is true; rather, it states that q is true when p and $p \to q$ are both true.) Furthermore, the proof of theorems in the if..., then... form reduces to the justification of a true conclusion given the fact that the hypothesis is true. It would be absurd to assume that we are starting with a false hypothesis; however, even in this case the logical argument provided by 1.2.2 still remains valid.

In general, a direct proof of an implication involves more than one step. Suppose that we wish to prove $p \to q$ and we know that $p \to r$

and $r \to q$ are true for some proposition r different from q. Can we conclude that $p \to q$ is true? The answer is yes, since

(1) $$[(p \to r) \wedge (r \to q)] \to (p \to q)$$

is a tautology (see 1.2.4). Suppose further that $p \to r_1$, $r_1 \to r_2$, $r_2 \to q$ are true. Can we conclude that $p \to q$ is true? Yes. $p \to r_2$ is true by one application of (1); and since $p \to r_2$ and $r_2 \to q$ are true, we conclude that $p \to q$ is true by applying (1) again. Repeated application of (1) allows us to use as many steps as necessary to prove a theorem in the if..., then... form.

To construct a truth table for the compound proposition (1) requires eight rows rather than four, since we have three simple components instead of two (see 1.2.4).

1.2.4 TABLE

p	q	r	$p \to r$	$r \to q$	$p \to q$	$(p \to r) \wedge (r \to q)$	$[(p \to r) \wedge (r \to q)] \to (p \to q)$
T	T	T	T	T	T	T	T
T	T	F	F	T	T	F	T
T	F	T	T	F	F	F	T
T	F	F	F	T	F	F	T
F	T	T	T	T	T	T	T
F	T	F	T	T	T	T	T
F	F	T	T	F	T	F	T
F	F	F	T	T	T	T	T

EXAMPLE. Let us give a proof and an analysis of the proof for the statement "if $2x - 4 = x - 1$ then $x = 3$."

Proof and analysis: We make the assumption that $2x - 4 = x - 1$ is true and deduce, using some properties of elementary algebra, that $2x - x = 4 - 1$ is true based on this assumption. By 1.2.2, "if $2x - 4 = x - 1$, then $2x - x = 4 - 1$" is true (this is one of the intermediate true conditionals $p \to r$). Next assuming $2x - x = 4 - 1$ is true, we deduce $x = 3$ is true based on this assumption. Again by 1.2.2, "if $2x - x = 4 - 1$, then $x = 3$" is true (this is another of the intermediate true conditionals $r \to q$). Now using the tautology (1), we conclude "if $2x - 4 = x - 1$, then $x = 3$" is true (this is $p \to q$). ▲

Although the problem of proving $p \to q$ in this example required three steps, notice that the technique of assuming p is true and deducing that q is true was still sufficient to conclude that $p \to q$ is true. This technique may be used in any proof as long as the number of steps remains finite. As for the proof of "if $2x - 4 = x - 1$, then $x = 3$," without the analysis we have

$$2x - 4 = x - 1,$$
$$2x - x = 4 - 1,$$
$$x = 3.$$

Clearly this result verifies the if..., then... statement as true, for we started with the assumed equality between $2x - 4$ and $x - 1$ and deduced that x is the number 3.

EXAMPLE. Prove that if x is an odd integer, then x^2 is an odd integer.

Proof: (Recall that x is an odd integer if x is of the form $2n + 1$ for some integer n.) Let x be an odd integer. Then $x = 2n + 1$ for some integer n and $x^2 = (2n + 1)^2 = 4n^2 + 4n + 1$. Now $4n^2 + 4n + 1 = 2(2n^2 + 2n) + 1 = 2k + 1$, where $k = 2n^2 + 2n$ is an integer. Since x^2 is of the form $2k + 1$, where k is an integer, we conclude that x^2 is an odd integer. (We assumed that the hypothesis is true and showed that the conclusion is true; thus the if..., then... statement is true.)▲

There are many paths a direct proof may follow. For this reason, no one formula or definite pattern will always produce a proof. The skill of building a chain of logically valid arguments develops only through insight and experience. There is no substitute for imitation and practice in learning this basic mathematical skill.

EXERCISES

1. Give an example of a conditional statement that has a different truth value from its converse.
2. Give an example of a conditional statement that has the same truth value as its converse.
†3. Explain why "if $2 + 1 = 4$, then $3 - 4 = 0$" and "if $2 + 1 = 4$, then $2 + 1 \neq 5$" are true conditional statements. (*Hint:* See 1.1.1.)

4. Give a proof and an analysis of the proof for $(\sqrt{x-2} = 4) \rightarrow x = 18$.

5. Prove that if x is an even integer, then x^2 is an even integer. (Recall that x is an even integer if x is of the form $2n$ for some integer n.)

6. Prove that if x and y are odd integers, then xy is an odd integer. (*Hint:* Pattern your proof after the last example in this section.)

†7. A good mathematics student profits as much from his mistakes as from his successes. Find the error in the following argument. If x is divisible by 4, then x is divisible by 2. Therefore, since 18 is divisible by 2, we know that 18 is divisible by 4.

†8. If the conditional statement $p \rightarrow q$ and its converse $q \rightarrow p$ are true, what can we say about the truth value of $p \leftrightarrow q$ (see Exercise 5 in 1.1)? Does this suggest an approach for proving $p \leftrightarrow q$?

9. Show that $(p \wedge \backslash q) \leftrightarrow \backslash (p \rightarrow q)$ is a tautology. With this result, state the negation of each of the following, using the conjunction.

   (a) If a is a negative integer, then $-a$ is a positive integer.

   (b) If a^2 is a positive integer, then a is a positive integer.

10. Construct a truth table for each of the following.

    (a) $(p \vee q) \vee r$.

    (b) $p \vee (q \vee r)$.

    (c) $[p \rightarrow (q \wedge r)] \leftrightarrow [(p \rightarrow q) \wedge (p \rightarrow r)]$.

1.3   INDIRECT PROOF AND THE CONTRAPOSITIVE

It is often quite difficult to construct a direct proof of an implication. In many such instances, the following tautology is helpful.

1.3.1 THEOREM. The compound proposition
$$(p \rightarrow q) \leftrightarrow (\backslash q \rightarrow \backslash p)$$
is a tautology.

Proof: Exercise 1 of this section.▲

The proposition $\backslash q \rightarrow \backslash p$ is called the *contrapositive* of $p \rightarrow q$. The logical equivalence (both have exactly the same truth table) of these two propositions suggests that a proof for either one suffices as a proof for the other.

EXAMPLE. Suppose that x is an integer and we wish to prove "if x^2 is an even integer, then x is an even integer."

In attempting a direct proof, we assume that x^2 is an even integer and write $x^2 = 2n$ for some integer n. Then $x = \sqrt{2n}$. What next? We have no way of concluding n is such that $\sqrt{2n}$ is an even integer, so no direct proof is in sight (at least not from this end of the pen).

Let us consider the contrapositive "if x is an odd integer (not an even integer), then x^2 is an odd integer." A direct proof of this statement is quite easy—see the last example in 1.2. Since the contrapositive is true, by 1.3.1, the implication "if x^2 is an even integer, then x is an even integer" is also true.

A direct proof for $\backslash q \rightarrow \backslash p$ is, by 1.3.1, also a proof for $p \rightarrow q$. Since it is not a direct proof for $p \rightarrow q$, it is called a *proof by contraposition*.

When attempts to prove $p \rightarrow q$ fail using both direct proof and proof by contraposition approaches, try an indirect proof—*a proof by contradiction*. This type of proof is based on the principle that sound reasoning produces a false conclusion only when the hypothesis is false. The most familiar type of indirect proof is based on the following tautology.

1.3.2 THEOREM. The compound proposition
$$[(p \wedge \backslash q) \rightarrow \backslash p] \leftrightarrow (p \rightarrow q)$$
is a tautology.

Proof: Exercise 4 of this section.▲

The propositions $(p \wedge \backslash q) \rightarrow \backslash p$ and $p \rightarrow q$ are logically equivalent; thus a proof for either one suffices as a proof for the other. According to 1.2.2, since $(p \wedge \backslash q) \rightarrow \backslash p$ is an if..., then... statement, we may attain a proof by assuming that $p \wedge \backslash q$ is true and deducing that $\backslash p$ is true. If we assume that $p \wedge \backslash q$ is true, then clearly p is true. Yet we are to deduce that $\backslash p$ is true—that is, we are to deduce a fallacy. Hence the name proof by contradiction.

A more usable form for an indirect proof of $p \rightarrow q$ is to assume $p \wedge \backslash q$ is true and then deduce a fallacy that is not necessarily the negation of the hypothesis p. Justification for this more general indirect proof approach is based on the following theorem.

1.3.3 THEOREM. If r is a true proposition, then
$$[(p \wedge \backslash q) \to \backslash r] \leftrightarrow (p \to q)$$
is a tautology.

Proof: The table in 1.3.4 establishes the result.▲

1.3.4 TABLE

p	q	r	$p \to q$	$p \wedge \backslash q$	$(p \wedge \backslash q) \to \backslash r$	$[(p \wedge \backslash q) \to \backslash r] \leftrightarrow (p \to q)$
T	T	T	T	F	T	T
T	F	T	F	T	F	T
F	T	T	T	F	T	T
F	F	T	T	F	T	T

EXAMPLE. We show that $\sqrt{2}$ is not a rational number by presenting the classical proof by contradiction.

Proof: Let p be the proposition "$x = \sqrt{2}$," q the proposition "$x \neq a/b$, where a and b are integers (i.e., x is not rational)," and r the proposition "a/b is in lowest terms." Clearly, if we assume that q is false (i.e., if we can choose integers a and b so that $x = a/b$), then we can choose a and b so that r is true.

We now establish $(p \wedge \backslash q) \to \backslash r$ by assuming that $p \wedge \backslash q$ is true and by deducing $\backslash r$. We have $x = \sqrt{2}$ and $x = a/b$, so $\sqrt{2} = a/b$. Hence $a = \sqrt{2}b$ and $a^2 = 2b^2$. a^2 is even, so, by the first example in this section, a is even and we can write $a = 2t$. $a = \sqrt{2}b$ and $a = 2t$; therefore $\sqrt{2}b = 2t$ and $b^2 = 2t^2$. Hence b is even. If a and b are both even, then a/b is not in lowest terms and we have deduced $\backslash r$. Finally, we conclude, via 1.3.3, that $p \to q$ is true; that is, if $x = \sqrt{2}$, then x is not a rational number.▲

EXERCISES

1. Construct a truth table for the tautology in 1.3.1.

†2. State the converse and contrapositive of each of the following:

 (a) $p \to q$,   (b) $q \to p$,

 (c) if lines L_1 and L_2 have no points in common, then they are parallel,

 (d) if a number ends with the digit 5, then the number is divisible by five.

3. Let x and y be integers. Give a proof by contraposition for "if xy is an odd integer, then x and y are odd integers."
4. Construct a truth table for the tautology in 1.3.2.
5. Show that $[(p \wedge \backslash q) \to q] \leftrightarrow (p \to q)$ is a tautology. This tautology gives rise to another frequently used indirect proof. Outline the procedure for giving an indirect proof using this tautology.
6. Distinguish between proof by contraposition and proof by contradiction.
†7. Show that $\backslash(p \wedge \backslash q) \leftrightarrow (p \to q)$ is a tautology.
†8. In view of Exercise 7, how are $p \wedge \backslash q$ and $p \to q$ related?

1.4   ADDITIONAL METHODS OF PROOF

Direct proof, proof by contraposition, and proof by contradiction are by far the most common methods of proof. However, several noncommon techniques are also used from time to time. This section illustrates two such techniques and concludes with a discussion of disproof and counterexample.

1.4.1 THEOREM. $[p \to (q \vee r)] \leftrightarrow [(p \wedge \backslash q) \to r]$ is a tautology.

Proof: Exercise 1 of this section.▲

EXAMPLE. Let a and b be real numbers. Prove that "if $ab = 0$, then $a = 0$ or $b = 0$."

Proof: Notice that the statement is in the form $p \to (q \vee r)$. By utilizing 1.4.1, if we assume that $p \wedge \backslash q$ is true and can deduce r, then the statement is true. To carry out the proof, let us assume that $ab = 0$ and $a \neq 0$. Then $1/a$ exists and

$$b = 1 \cdot b = \left(\frac{1}{a} \cdot a\right) b = \frac{1}{a}(ab) = \frac{1}{a} \cdot 0 = 0.$$

Hence $b = 0$ (i.e., we have deduced r) and the statement is proved.▲

At first glance we might think that the proof is not complete until $(p \wedge \backslash r) \to q$ is also verified. This reasoning is a common error, for clearly, according to 1.4.1, the proof is complete.

1.4.2 THEOREM. $[(p \to q) \land (\backslash p \to q)] \leftrightarrow q$ is a tautology.

Proof: Exercise 4 of this section.▲

EXAMPLE. Let n be an integer. Show that $n^2 - n$ is an even integer.

Proof: Let q be the statement "$n^2 - n$ is an even integer" and p the statement "n is an even integer." To show that q is true, we need to show that both $p \to q$ and $\backslash p \to q$ are true.

Case 1. Assume that n is even. Then $n = 2k$ for some integer k and
$$n^2 - n = (2k)^2 - 2k = 2(2k^2 - k).$$
Hence $n^2 - n$ is even.

Case 2. Assume that n is odd. Then $n = 2m + 1$ for some integer m and
$$n^2 - n = (2m + 1)^2 - (2m + 1) = 2(2m^2 + m).$$
Hence $n^2 - n$ is even.

By 1.4.2, q is true, since Case 1 establishes $p \to q$ and Case 2 establishes $\backslash p \to q$.▲

To say the proposition $p \to q$ is true means that every instance of the statement is true. In particular, each time p is true, q must also be true. Therefore, if there is at least one instance in which p is true and q is false, we must conclude that $p \to q$ is false. Such an instance is called a *counterexample* for $p \to q$. Obviously if a counterexample for $p \to q$ exists, then $p \to q$ is false—thus the phrase "disproof by counterexample."

EXAMPLE. Disprove the proposition "if x is an odd integer, then x^2 is even."

The instance when $x = 3$ produces a true hypothesis (3 is an odd integer) and a false conclusion (3^2 is even). Therefore this instance serves as a counterexample, and the proposition is false. Notice that many instances can serve as counterexamples.

A general proposition p is true if and only if every instance is true. Hence one counterexample acts as a disproof of the proposition.

EXAMPLE. Disprove the statement "2n $>$ n for each integer n."

We might consider the statement to be true if we were only familiar with the positive integers, since each positive integer produces a true instance. However, taking $n = -1$ produces a counterexample; therefore the statement is false.

It is important to note that although there are infinitely many true instances, the proposition is false, since there is at least one false instance.

EXERCISES

1. Construct a truth table to show that the compound proposition in 1.4.1 is a tautology. (Recall that the table will have eight rows.)

†2. Let x and y be integers. Prove that "if xy is an even integer, then x is even or y is even" using 1.4.1.

3. Show that $[p \to (q \wedge r)] \leftrightarrow [(p \to q) \wedge (p \to r)]$ is a tautology. This tautology suggests that we should establish both $p \to q$ and $p \to r$ to prove $p \to (q \wedge r)$.

4. Construct a truth table for the compound proposition in 1.4.2 to show that it is a tautology.

5. Assuming the necessary properties of $<$ (is less than) on the set of integers, prove "$a^2 > 0$ for each integer a" using 1.4.2. (*Hint:* Take q to be "$a^2 > 0$" and p to be "$a < 0$.")

6. Prove that $n^2 + n$ is an even integer for each integer n.

†7. Disprove each of the following.

   (a) If x is an even integer, then x^2 is odd.

   (b) If $x^2 = 9$, then $x = -3$.

   (c) $x^2 + 4 \neq (x + 2)^2$ for all integers x.

1.5   SETS

It is disconcerting to some people to learn that a mathematician makes no attempt to define certain basic concepts and also blithely assumes true truth values for a number of propositions that he calls axioms. Actually, mathematics in this aspect is no different from many other disciplines. Raw material is necessary before any construction can take place. The mathematician's raw materials are these undefined terms and axioms. It is in terms of these "basic truths" that definitions and theorems are formed.

In this text, we will not define the term "set." We will assume that sets exist and are collections of things called *elements* or *members*. We will also assume the basic properties of the integers, the rational numbers, and the real numbers.

Notationally, capital letters will be used to represent sets, lower case letters to represent elements, and the symbol $\in$ to denote the relationship of membership. We read $a \in A$ as "a is an element of the set A" or, more informally, "a belongs to A." The negation of $a \in A$, denoted $a \notin A$, is read "a is not an element of A" or "a does not belong to A." To illustrate, let X denote the set of letters of the alphabet. Then $a \in X$ and $b \in X$; however, 2 is not a letter of the alphabet so we write $2 \notin X$.

There is one basic qualification that sets must satisfy in order to be mathematically acceptable—they must be well defined. We will restrict our consideration to well-defined sets.

1.5.1 DEFINITION. A set A is *well defined* if the statement $x \in A$ is a proposition for each object x.

Since propositions are either true or false, a set is well defined only when $x \in A$ or $x \notin A$ for each replacement of x. No judgment is permitted with propositions; therefore none is permitted with well-defined sets. The set of pink roses, for example, is not a well-defined set, whereas the set of positive integers is. Since consideration is to be restricted to well-defined sets, we adopt the convention that "set" will be interpreted as "well-defined set."

There are three basic ways of describing a set. The first is to list all elements of the set between braces. For example,

$$\{1,2,3,4,5\} \quad \text{and} \quad \{a,b,c\}.$$

Clearly this method is only convenient for sets with relatively few members.

The second technique is to establish a pattern. For example,

$$\{1,2,3,4,\ldots,10\}, \quad \{2,4,6,\ldots\}, \quad \text{and} \quad \{\ldots,-3,-2,-1,0,1,2,3,\ldots\}.$$

The first set consists of all integers from 1 to 10 inclusive, the second consists of all positive even integers, and the third is the set of all integers.

Some particular relationship between successive elements is necessary to utilize this technique.

The third way of describing a set is by using a statement that is true for each element of the set and false for every object not belonging to the set. For example,

$$\{x \mid x \text{ is an integer between 0 and 11}\},$$
$$\{y \mid y \text{ is a positive even integer}\},$$
$$\{z \mid z \text{ is an integer}\}.$$

These sets contain, respectively, the same elements as those sets illustrating the pattern technique. The vertical bar | is read "having the property that" or, more informally, "such that." Therefore we read the first set as "the set of all x having the property that x is an integer between 0 and 11."

The method used to describe a set does not change the set. Consequently, we will always try to use the descriptive method that best fits a given situation. The following sequence of definitions and examples initiates the development of the elementary set theory needed throughout this text.

1.5.2 DEFINITION. The set U consisting of all elements under consideration at a given time is called the *universal set*.

All sets referred to hereafter will be subsets of U in the following sense.

1.5.3 DEFINITION. Let A and B be sets. If each element of A is also an element of B, then A is a *subset* of B and we write $A \subseteq B$ or $B \supseteq A$.

Informally, we read $A \subseteq B$ as "A is contained in B" and $B \supseteq A$ as "B contains A."

1.5.4 DEFINITION. Let A and B be sets. If $A \subseteq B$ and $B \subseteq A$, then A and B are *equal*, and we write $A = B$.

Since the familiar number systems will be referred to extensively, we pause at this time and assign special symbols to them.

	Integers	Rational Numbers	Real Numbers
All	Z	Q	F
Positive	Z_+	Q_+	F_+
Nonzero	Z_0	Q_0	F_0

From the chart we see that Z_0 represents the set of all nonzero integers, Q_+ the set of all positive rational numbers, and F the set of real numbers.

EXAMPLES

1. Relative to the preceding chart, the universal set is F, and we have $Z_+ \subseteq Z_0, Z_0 \subseteq Z, Z \subseteq Q, Q_+ \subseteq Q, Q_+ \subseteq F_+$, etc.
2. Let $A = \{$one, two, three$\}$ and $B = \{1,2,3\}$. Then $A = B$.
3. Let $D = \{4,5,6,7\}$. If $E = \{5,7,6,4\}$, then $D = E$.
4. Let $H = \{1,2\}$. If $G = \{1,2,1\}$, then $H = G$.

Example 2 illustrates the fact that the symbols used to represent the elements of a set do not alter or change the set. Example 3 indicates that the order in which elements are listed in a set is irrelevant. The membership of the set is unchanged. The membership of a set also remains unchanged no matter how many times an element appears symbolically in the set. This point is illustrated by Example 4.

For a given set A, clearly $A \subseteq A$. However, A, as a subset of itself, is different from all other subsets of A. The next definition points out this difference.

1.5.5 DEFINITION. Let B be a set. If $A \subseteq B$ and $A \neq B$, then A is a *proper subset* of B and we write $A \subset B$ or $B \supset A$.

Read $A \subset B$ as "A is properly contained in B" and $B \supset A$ as "B properly contains A." If A is not a (proper) subset of B, we write $(A \nsubseteq B) A \nsubseteq B$. Clearly, for any set X, we have $X \subseteq X$ and $X \nsubseteq X$. To show that a set S is a proper subset of T, we must establish that $S \subseteq T$ and then obtain an element $t \in T$ such that $t \notin S$.

1.5.6 DEFINITION. The *empty set* or *null set*, denoted by $\emptyset$, is the set that has no elements.

The empty set arises quite naturally in many ways as we shall see in Section 1.6. Also, it can be characterized in a number of ways. For example, $\{x \mid x$ is an integer between 0 and 1$\}$ has no elements; thus it is the empty set. On the other hand, $\{0\}$ is not the empty set, since it contains the element 0. Is $\{\emptyset\}$ empty?

EXERCISES

1. Describe each of the following sets by employing a method different from that used.
   (a) $\{-3, -2, -1, 0, 1, 2, 3\}$ (b) $\{1, \frac{1}{2}, \frac{1}{3}, \ldots\}$
   (c) $\{x \mid x \in Z$ and $x > 5\}$ (d) $\{x \mid x \in Z$ and $x^2 - 1 = 0\}$

2. List all subsets of $\{1, 2, 3\}$.

†3. Classify each of the following as true or false:
   (a) $a \in \{a, b\}$ (b) $\{b\} \in \{a, b\}$
   (c) $\emptyset \subset \{a, b\}$ (d) $\emptyset \in \{a, b\}$

4. Why is $\emptyset$ a well-defined set? (Answer the question relative to 1.5.1.)

†5. Using the technique of proof by contradiction, prove that the empty set is a subset of every set. Conclude that there is only one empty set.

6. Is it true that $\emptyset$ is a proper subset of every set? If your answer is no, then justify it with a counterexample.

†7. Prove that if $A \subseteq B$ and $B \subseteq C$, then $A \subseteq C$. (*Hint:* Use an element argument. Assume that $A \subseteq B$ and $B \subseteq C$, and show that if $x \in A$, then $x \in C$.)

1.6   COMBINATIONS OF SETS

Just as propositions can be combined to form new propositions, sets can be combined to form new sets. The several basic ways that we will use to combine sets are given in

1.6.1 DEFINITION. Let A and B be sets.

1. The *union* of A and B, denoted $A \cup B$, is the set of all elements that belong to either A or B or to both A and B. Symbolically,

$$A \cup B = \{x \mid x \in A \text{ or } x \in B\}.$$

2. The *intersection* of A and B, denoted $A \cap B$, is the set of all elements that belong to both A and B. Symbolically,

$$A \cap B = \{x \mid x \in A \text{ and } x \in B\}.$$

3. *The difference* of A and B, denoted $A \setminus B$, is the set of elements that belong to A but not to B. Symbolically,

$$A \setminus B = \{x \mid x \in A \text{ and } x \notin B\}.$$

EXAMPLES

1. Let $A = \{1,2,3,4,5,6\}$ and $B = \{4,6,8,10\}$. Then $A \cup B = \{1,2,3,4,5,6,8,10\}$, $A \cap B = \{4,6\}$, and $A \setminus B = \{1,2,3,5\}$. Notice that $A \cup B = B \cup A$ and $A \cap B = B \cap A$, but $A \setminus B \neq B \setminus A$.

2. $Q \cup F = F$, $Q \cap F = Q$, $Q \setminus F = \emptyset$, and $F \setminus Q$ is the set of all irrational numbers.

A *venn diagram* (named after the nineteenth-century English logician John Venn) is a mnemonic device that is exceptionally useful from an intuitive point of view when combinations of sets are considered. It pictures the universal set U as a rectangle and its interior. Particular subsets of U are pictured as circles and their interiors. For example, the venn diagrams in 1.6.2 exhibit A, B, C, and D as subsets of U. They

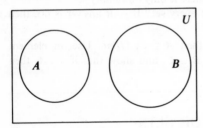
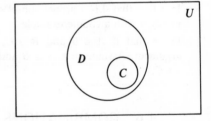

1.6.2 FIGURE

further indicate that $A \cap B = \emptyset$ and $C \subset D$. In 1.6.3., the definitions of union and intersection are represented pictorially by the shaded areas.

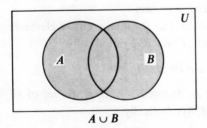
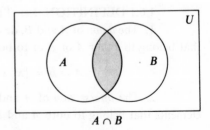

$A \cup B$ $A \cap B$

1.6.3 FIGURE

The difference of U and B is called the *complement* of B and is usually denoted by $\setminus B$ rather than by $U \setminus B$. Also, $A \setminus B$ is sometimes called the complement of B in A. The shaded areas in 1.6.4 represent pictorally $A \setminus B$ and $\setminus B$.

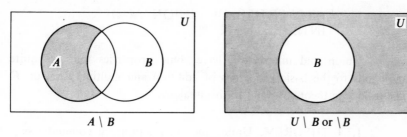

$A \setminus B$ $U \setminus B$ or $\setminus B$

1.6.4 FIGURE

We mentioned in Section 1.5 that the empty set arises in many ways. One natural way is to intersect A and B when they have no elements in common ($A \cap B = \emptyset$). Pairs of sets having this property are called *disjoint sets*.

EXERCISES

1. Let $U = \{1,2,3,\ldots,10\}$, $A = \{1,2,5,9,10\}$, $B = \{3,4,5\}$, and $C = \{5,6,8,10\}$. Calculate

   (a) $A \cup B$,   (b) $(A \cup B) \cup C$,   (c) $A \cap B$,   (d) $(A \cap B) \cap C$.

†2. Using U, A, B, and C as in Exercise 1, find

   (a) $\setminus(A \cup B)$,   (b) $\setminus A \cap \setminus B$,   (c) $\setminus(A \cap B)$,   (d) $\setminus A \cup \setminus B$.

3. Use a venn diagram to show that

   $\setminus(A \cup B) = \setminus A \cap \setminus B$ and $\setminus(A \cap B) = \setminus A \cup \setminus B$.

4. Show by counterexample that

$$(A \setminus B) \setminus C \neq A \setminus (B \setminus C).$$

5. Use venn diagrams to establish each of the following:

   (a) $A \setminus B = A \cap \setminus B$,   (b) $(A \cap B) \cup (A \cap \setminus B) = A$.

†6. What conditions must A and/or B satisfy if

   (a) $A \cap \emptyset = A$,   (b) $A \cup \emptyset = \emptyset$,   (c) $A \cap B = A$,

   (d) $A \cup B = A$,   (e) $A \setminus B = A$,   (f) $A \setminus B = \emptyset$,

             (g) $A \setminus B = B$,   (h) $\setminus A = \emptyset$.

7. Prove each of the following, using the definitions we have given.

   (a) If $A \subseteq B$, then $\setminus B \subseteq \setminus A$.

   (b) $A \cap B \subseteq A$.

(c) If $A \subseteq B$, then $A \setminus B = \emptyset$.

(d) $A \subseteq B$ if and only if $A \cup B = B$.

(e) If A and B are disjoint, then $A \setminus B = A$.

1.7   PROPERTIES OF UNION AND INTERSECTION

Union and intersection have some properties that are quite analogous to the basic properties of addition and multiplication on F. The following theorems point to this analogy.

1.7.1 THEOREM. Union and intersection are commutative— that is, for any sets A and B we have
$$A \cup B = B \cup A,$$
$$A \cap B = B \cap A.$$

Proof: Immediate from the definition of union and intersection. ▲

1.7.2 THEOREM. Union and intersection are associative—that is, for any sets A, B, and C we have
$$(A \cup B) \cup C = A \cup (B \cup C),$$
and
$$(A \cap B) \cap C = A \cap (B \cap C).$$

Proof: Exercise 1 of this section. ▲

1.7.3 THEOREM. Intersection distributes over union—that is, for any sets A, B, and C we have
$$A \cap (B \cup C) = (A \cap B) \cup (A \cap C).$$

Proof: According to 1.5.4, we need to establish containment in both directions. We first verify that $A \cap (B \cup C) \subseteq (A \cap B) \cup (A \cap C)$. Let $x \in A \cap (B \cup C)$; then $x \in A$ and $x \in B \cup C$ by 1.6.1.2. Now $x \in B \cup C$ implies $x \in B$ or $x \in C$. First, suppose that $x \in B$; then $x \in A$ and $x \in B$. Therefore $x \in A \cap B$ and clearly $x \in (A \cap B) \cup (A \cap C)$. For the second case—namely, $x \in C$—we have $x \in A \cap C$ and clearly $x \in (A \cap B) \cup (A \cap C)$. Thus in both cases $x \in (A \cap B) \cup (A \cap C)$, whence $A \cap (B \cup C) \subseteq (A \cap B) \cup (A \cap C)$.

To verify $(A \cap B) \cup (A \cap C) \subseteq A \cap (B \cup C)$, let $y \in (A \cap B)$ $\cup (A \cap C)$. Then $y \in A \cap B$ or $y \in A \cap C$. Again we consider two cases. First, suppose that $y \in A \cap B$; then $y \in A$ and $y \in B$. Thus $y \in B \cup C$. Now $y \in A$ and $y \in B \cup C$ implies $y \in A \cap (B \cup C)$. Similarly, $y \in A \cap C$ implies $y \in A \cap (B \cup C)$. Thus in both cases $y \in A \cap (B \cup C)$, and we conclude that $(A \cap B) \cup (A \cap C) \subseteq A \cap (B \cup C)$. Finally, containment in both directions give equality. ▲

To further point to the analogy between the "algebra of sets" and the algebra of real numbers, notice the elementary facts displayed in 1.7.4. There is a similarity between the roles of union and addition, $\emptyset$ and 0, intersection and multiplication, U and 1, and subtraction and set difference.

1.7.4 TABLE

For Sets	For Real Numbers
$A \cup \emptyset = A$	$a + 0 = a$
$A \cap U = A$	$a \cdot 1 = a$
$A \cap \emptyset = \emptyset$	$a \cdot 0 = 0$
$A \setminus A = \emptyset$	$a - a = 0$
$A \setminus \emptyset = A$	$a - 0 = a$

In F, multiplication distributes over addition $[a(b + c) = ab + ac]$; however, addition does not distribute over multiplication $[a + (bc) \neq (a + b) \cdot (a + c)]$. The next result provides one example of how the algebra of sets differs from the algebra of real numbers.

1.7.5 THEOREM. Union distributes over intersection—that is, for any sets A, B, and C we have
$$A \cup (B \cap C) = (A \cup B) \cap (A \cup C).$$

Proof: Exercise 3 of this section. ▲

There is an abundance of additional properties that union and intersection satisfy, some of which are given in the exercises for this section.

Frequently in mathematics we need to consider a collection of sets (a set whose elements are sets). For such a collection, it is usually convenient to use the elements of some particular set as subscripts to label the sets in the collection. A nonempty set A is an *index set* for a collection $\mathscr{C}$ of sets if for each $a \in A$ there is a set S_a in $\mathscr{C}$. The set A can be any set—finite or infinite. Generally A is chosen to be a subset of Z_+. However, we cannot require A to be a subset of Z_+, since in some instances there are not enough positive integers to index a given collection of sets.

EXAMPLES

1. Let $A = \{1,2,\ldots,10\}$. Let $S_1 = \{-1,0,1\}$, $S_2 = \{-2,0,2\}$, $\ldots$, $S_{10} = \{-10,0,10\}$. Now $\mathscr{C} = \{S_i \mid i \in A\}$ is a collection of sets using A as an index set.

2. Let $M_n = \{x \mid x \in F \text{ and } -1/n < x < 1/n\}$ for n a positive integer. $\mathscr{F} = \{M_n \mid n \in Z_+\}$ is a collection of sets using Z_+ as an index set.

1.7.6 DEFINITION. Let $\mathscr{C} = \{S_a \mid a \in A\}$ be a collection of sets.
1. The *union* of $\mathscr{C}$ is the set
$$\cup \mathscr{C} = \{x \mid x \in S_a \text{ for at least one } a \in A\}.$$
2. The *intersection* of $\mathscr{C}$ is the set
$$\cap \mathscr{C} = \{x \mid x \in S_a \text{ for every } a \in A\}.$$

In Example 1, $\cup \mathscr{C} = \{-10,-9,\ldots,\ -1,0,1,\ldots,\ 10\}$ and $\cap \mathscr{C} = \{0\}$. In Example 2, $\cup \mathscr{F} = M_1$ because $M_i \subset M_1$ for each positive integer i greater than 1. Exercise 6 establishes that $\cap \mathscr{F} = \{0\}$.

EXERCISES

1. Prove 1.7.2.
2. Show that for each set A, $A \cup A = A$ and $A \cap A = A$. Do addition and multiplication on F have properties similar to these?
3. Prove 1.7.5.
†4. Prove each of the following:
    (a) If $A \subseteq B$, then $A \cup C \subseteq B \cup C$ and $A \cap C \subseteq B \cap C$.
    (b) If $A \subseteq C$ and $B \subseteq C$, then $A \cup B \subseteq C$.
    (c) Let $A \cup B = C$ and $A \cap B = \emptyset$, then $A = C \setminus B$.

5. Let A and B be subsets of C. Verify that
   (a) $C \setminus (A \cup B) = (C \setminus A) \cap (C \setminus B)$
   (b) $C \setminus (A \cap B) = (C \setminus A) \cup (C \setminus B)$

†6. Show that $\cap \mathscr{F} = \{0\}$ when $\mathscr{F}$ is the collection of sets given in Example 2 of this section.

7. Let S be any set. The *power set* of S is the collection of all subsets of S. Notationally, $P(S) = \{X \mid X \subseteq S\}$. If $T = \{1,2,3\}$, construct $P(T)$ and find $\cup P(T)$ and $\cap P(T)$.

8. Let S be a nonempty set. Show that $\cup P(S) = S$ and $\cap P(S) = \emptyset$ (see Exercise 7).

CHAPTER 2

BASIC CONCEPTS IN THE DEVELOPMENT OF ALGEBRAIC SYSTEMS

2.1 RELATIONS

When we see the word relation, we automatically think of parents, brothers, sisters, aunts, uncles, and so on. The idea of "being related" is well known to us. We can easily write down pairs of relatives in this rather intuitive sense. Mathematically, this is precisely what we do, and then we proceed to study the properties of these pairs.

The order in which a pair of relatives is written is extremely important. Consider the relation "is a brother of," and let Jim and Sally be children in the same family. Corresponding to the pair (Jim, Sally) we can write "Jim is a brother of Sally" but clearly we should

27

not write "Sally is a brother of Jim," which would correspond to the pair (Sally, Jim). Since the only difference between these two pairs is the order in which Jim and Sally are written, our first mathematical detail in the study of relations is the concept of *ordered pair* rather than simply "pair."

The ordered pair of elements a and b will be denoted by (a,b). Furthermore, we agree that $(a,b) = (c,d)$ if and only if $a = c$ and $b = d$.

2.1.1 DEFINITION. For sets A and B, the *cartesian product set*, $A \times B$, is the collection of all ordered pairs (a,b), where $a \in A$ and $b \in B$. Symbolically, $A \times B = \{(a,b) \mid a \in A \text{ and } b \in B\}$.

EXAMPLE. Let $A = \{1,2,3\}$ and $B = \{2,m\}$. Then we have
$A \times B = \{(1,2),(1,m),(2,2),(2,m),(3,2),(3,m)\}$ and
$B \times A = \{(2,1),(2,2),(2,3),(m,1),(m,2),(m,3)\}$.

We note that $(2,2) \in A \times B$ and $(2,2) \in B \times A$, but none of the other ordered pairs has this property. In particular, $(1,m) \in A \times B$ but $(1,m) \notin B \times A$; thus we conclude that $A \times B \neq B \times A$.

2.1.2 DEFINITION. A *relation from A to B* is any subset of $A \times B$. In the particular case where $A = B$, a relation will be called a relation on (in) A.

To see that this definition meets with our usual ideas about the nature of relations, consider the following.

EXAMPLES

1. "Is the mother of" is a relation from the set W of all women to the set P of all people. Some females will not occur as a first element of any of the pairs, while others will occur as first elements several times. If Mary has two children, Dave and Judy, then the ordered pairs (Mary,Dave) and (Mary,Judy) both belong to this relation. The set of all such ordered pairs is clearly a subset of $W \times P$.

2. The relation "is less than" on the set $S = \{1,2,3,4\}$ is the collection $\{(1,2),(1,3),(1,4),(2,3),(2,4),(3,4)\}$. Clearly we have a subset of $S \times S$.

3. $\{(x,1),(x,2),(z,1),(z,3)\}$ is a relation from S to T, where $S = \{x,y,z\}$ and $T = \{1,2,3,4\}$.

4. $\emptyset$ is a subset of $A \times B$ for any pair of sets A and B and can, therefore, be regarded as a relation from A to B (or as a relation on A since $\emptyset \subseteq A \times A$). We call $\emptyset$ the *empty relation*, since none of the elements of A and B is related by $\emptyset$.

5. For sets A and B, $A \times B \subseteq A \times B$, so $A \times B$ itself can be considered as a relation from A to B. We call this the *trivial relation* from A to B (every element of A is related to every element of B).

The common relations, such as equality ($=$), is less than ($<$), is the mother of, and so forth, suggest the following notation. If R is a relation from A to B, we write aRb if $(a,b) \in R$ and $a\mathrel{R\!\!\!/}b$ if $(a,b) \notin R$. Read aRb as "a is R-related to b" and $a\mathrel{R\!\!\!/}b$ as "a is not R-related to b." For equality on the set of rational numbers, for instance, we would be more likely to write $\frac{4}{2} = 2$ than $(\frac{4}{2},2) \in =$.

Many instances of a relation on a set arise throughout the study of mathematics. There are three important properties that such relations might have.

2.1.3 DEFINITION. Let R be a relation on a set A.

1. R is *reflexive* in A if aRa for each $a \in A$; that is, if $(a,a) \in R$ for each $a \in A$.

2. R is *symmetric* in A if aRb implies bRa; that is, $(b,a) \in R$ whenever $(a,b) \in R$.

3. R is *transitive* in A if aRb and bRc together imply aRc; that is, $(a,c) \in R$ whenever (a,b) and (b,c) are both in R.

EXAMPLES

1. Granting that a line is parallel to itself, the relation "is parallel to" on the set of all straight lines in a plane is reflexive, symmetric, and transitive.

2. The relation "is perpendicular to" on the same set as in Example 1 is symmetric but neither reflexive nor transitive.

3. The relation $R = \{(1,1),(1,2),(2,1),(3,3)\}$ on $S = \{1,2,3\}$ is symmetric but neither transitive nor reflexive in S, since $(2,2) \notin R$.

There is an important distinction between the reflexive property and the other two properties. The reflexive property states that each element belongs to at least one ordered pair in the relation. The symmetric and transitive properties are implications and can thus be true with no guarantee that any element belongs to an ordered pair in the relation.

Consider the following argument:

Let R be a relation on a set S that is both symmetric and transitive. For $a,b \in S$, if aRb, then because of symmetry we have bRa. Now with aRb, bRa, and transitivity, we must also have aRa.

It is possible that, without making the foregoing distinction, we could be led to believe that any relation that is both symmetric and transitive is also reflexive. Symmetry and transitivity, however, do not guarantee that a given element $a \in S$ is related to anything. In particular, we have no guarantee that $b \in S$ exists such that aRb. Thus we cannot conclude that each $a \in S$ is related to itself. Exercise 3, part (e), asks for an example of such a relation.

EXERCISES

1. Let $f \subseteq F \times F$ be defined by $f = \{(x,y)|x \in F$ and $y = x^2\}$.
    (a) Represent f graphically.
    (b) Is $(1,1) \in f$? Is f reflexive?

†2. Let $A = \{a \in F|0 \le a \le 1\}$ and $B = \{b \in F|0 \le b \le 2\}$.
    (a) Represent $A \times B$ graphically.
    (b) Represent the relation $\le$ from A to B graphically.

3. Show that reflexivity r, symmetry s, and transitivity t are independent properties of relations by giving an example (either mathematical or non-mathematical) of a relation on a set that satisfies each of the following:
    (a) $r, s, t,$ (b) $r, s, \backslash t,$ (c) $r, \backslash s, \backslash t,$
    (d) $r, \backslash s, t,$ (e) $\backslash r, s, t,$ (f) $\backslash r, s, \backslash t,$
                (g) $\backslash r, \backslash s, t,$ (h) $\backslash r, \backslash s, \backslash t.$

4. Repeat Exercise 3 with the added requirement that all the relations be on the set $A = \{1,2,3\}$.

†5. Classify the following relations on Z via reflexivity, symmetry, and transitivity.
    (a) xRy if and only if $x - y$ is an even integer.
    (b) xRy if and only if $x + y$ is an odd integer.

(c) xRy if and only if x and y have the same remainder when divided by 5.

(d) xRy if and only if $x < y$.

6. Given nonempty sets A and B, prove that $A \times B = B \times A$ if and only if $A = B$.

2.2 EQUIVALENCE RELATIONS AND PARTITIONS

A relation on a set A that decomposes A into a collection of subsets whose union is A belongs to a significant class of relations. Such relations, called *equivalence relations*, play a unifying role in mathematics, since they are basic to all areas of mathematics.

2.2.1 DEFINITION. A relation $\sim$ (tilde) on a set A is an *equivalence relation* on A if $\sim$ is reflexive, symmetric, and transitive on A. Read $a \sim b$ as "a is equivalent to b" or as "a tilde b."

In order to examine the nature of an equivalence relation on A, we first define the subsets of importance.

2.2.2 DEFINITION. Let $\sim$ be an equivalence relation on A. For each $a \in A$, we denote by $\bar{a}$ the collection of all $\sim$ relatives of a. Symbolically, $\bar{a} = \{x \in A \mid x \sim a\}$. $\bar{a}$ is called the *equivalence class* of $\sim$ represented (determined) by a.

EXAMPLE. The relation "is in the same grade as" on the set of Jones County Public School pupils (grades 1 through 12) is an equivalence relation. If Johnny is a third grader, then the equivalence class determined by Johnny is the set of all pupils in the third grade. How many equivalence classes are there? Do any two equivalence classes have any pupils in common? What can you say about the equivalence class determined by Johnny and the equivalence class determined by Betty, who is also a third grader?

2.2.3 LEMMA. Let $\sim$ be an equivalence relation on a set A. If $a,b \in A$ such that $a \sim b$, then $\bar{a} = \bar{b}$ (i.e., the representative a of the equivalence class $\bar{a}$ is not necessarily unique).

Proof: We show that $\bar{a} = \bar{b}$ by establishing $\bar{a} \subseteq \bar{b}$ and $\bar{b} \subseteq \bar{a}$. Let $x \in \bar{a}$. Then $x \sim a$. Since $\sim$ is transitive, $x \sim a$ and $a \sim b$ imply $x \sim b$; that is, $x \in \bar{b}$ and we conclude that $\bar{a} \subseteq \bar{b}$. Similarly, $\bar{b} \subseteq \bar{a}$. ▲

2.2.4 COROLLARY. Let $\sim$ be an equivalence relation on A. For equivalence classes $\bar{a}$ and $\bar{b}$, either $\bar{a} = \bar{b}$ or $\bar{a} \cap \bar{b} = \emptyset$.

Proof: Suppose that $\bar{a} \cap \bar{b} \neq \emptyset$. Then there is an element x so that $x \in \bar{a}$ and $x \in \bar{b}$. By 2.2.3, $\bar{a} = \bar{x}$ and $\bar{b} = \bar{x}$. Therefore $\bar{a} = \bar{b}$.

Equivalence classes possess three important properties. We formalize them in the following definition.

2.2.5 DEFINITION. Let A be a nonempty set. A *partition* of A is a collection $\mathscr{P}$ of subsets of A with the following properties:
1. Each $X \in \mathscr{P}$ is nonempty,
2. If $X, Y \in \mathscr{P}$ and $X \neq Y$, then $X \cap Y = \emptyset$, and
3. $\cup \mathscr{P} = A$.

A collection of sets satisfying property (2) is said to be *pairwise disjoint*. A partition, then, of a nonempty set A is a collection of pairwise disjoint nonempty subsets of A whose union is A. The elements of a partition are called *cells*.

A good intuitive idea of a partition is given by a jigsaw puzzle. Each piece of the puzzle represents a cell of the partition and the whole picture represents the set A. The required three properties of a partition are easily seen to be satisfied: (1) each piece contains some of the picture and thus is nonempty, (2) different pieces might interlock but they do not overlap, and (3) the union of the pieces is the whole puzzle.

The next theorem shows the correspondence between the set of all equivalence relations on A and the set of all partitions of A.

2.2.6 THEOREM. Let A be a nonempty set. Each equivalence relation on A determines a partition of A, and, conversely, each partition of A determines an equivalence relation on A.

Proof: Let $\sim$ be an equivalence relation on A and let $\mathscr{P}$ be the collection of equivalence classes determined by $\sim$. Recall that $\bar{a} = \{x \in A \mid x \sim a\}$; thus $\mathscr{P} = \{\bar{a} \mid a \in A\}$.

1. For each $\bar{a} \in \mathscr{P}$, $a \in \bar{a}$, since $\sim$ is reflexive. Therefore each cell in $\mathscr{P}$ is a nonempty subset of A.

2. By 2.2.4, the elements of $\mathscr{P}$ are pairwise disjoint.

3. Since $a \in \bar{a}$ for each $a \in A$, we have $A \subseteq \cup \mathscr{P}$ and clearly $\cup \mathscr{P} \subseteq A$. Therefore $\cup \mathscr{P} = A$.

By (1), (2), and (3), $\mathscr{P}$ is a partition of A.

Conversely, let $\mathscr{P}$ be a partition of A. Define $\sim$ on A by $x \sim y$ if and only if x and y belong to the same cell in $\mathscr{P}$.

1. Since $\cup \mathscr{P} = A$, each $a \in A$ is in some cell and clearly is in the same cell with itself. Hence $a \sim a$ for all $a \in A$ and $\sim$ is reflexive.

2. Let $a \sim b$. Then a and b are in the same cell, which certainly implies b and a are in the same cell; that is, $b \sim a$. Hence $\sim$ is symmetric.

3. Let $a \sim b$ and $b \sim c$. Then a and b are in the same cell and so are b and c. Different cells are disjoint; so a and c must be in the same cell, which implies $a \sim c$. Hence $\sim$ is transitive.

By (1), (2), and (3), $\sim$ is an equivalence relation on A.▲

EXAMPLE. List all equivalence relations on $A = \{a,b,c\}$. We first list all partitions on A and then obtain the associated equivalence relations on A.

Partitions	Equivalence Relations
$\mathscr{P}_1 = \{A\}$	$R_1 = \{(a,a),(b,b),(c,c),(a,b),(b,a),$
	$\quad (a,c),(c,a),(b,c),(c,b)\}$
$\mathscr{P}_2 = \{\{a\},\{b,c\}\}$	$R_2 = \{(a,a),(b,b),(c,c),(b,c),(c,b)\}$
$\mathscr{P}_3 = \{\{a,b\},\{c\}\}$	$R_3 = \{(a,a),(b,b),(c,c),(a,b),(b,a)\}$
$\mathscr{P}_4 = \{\{a\},\{b\},\{c\}\}$	$R_4 = \{(a,a),(b,b),(c,c)\}$
$\mathscr{P}_5 = \{\{a,c\},\{b\}\}$	$R_5 = \{(a,a),(b,b),(c,c),(a,c),(c,a)\}$

Are these all of the equivalence relations on A? Why?

EXERCISES

1. $\mathscr{P} = \{\{1,2,4\},\{3,5\}\}$ is a partition of $A = \{1,2,3,4,5\}$. Let R be the equivalence relation on A induced by $\mathscr{P}$. List the elements of R.

†2. Let $A = \{1,2,3\}$ and $R = \{(1,1),(2,2),(2,3)\}$.

    (a) Add the minimum number of ordered pairs to R in order to make R an equivalence relation on A.

    (b) Give the partition of A induced by this equivalence relation on A.

3. In 2.2.3, verify in detail that $\bar{b} \subseteq \bar{a}$.

†4. In 2.2.3, we established that equivalence class representatives are not necessarily unique. What can be said about equivalence classes with unique representatives? Give an example of an equivalence relation such that every equivalence class has this property.

5. Exercise 5, part (c), of Section 2.1 gives an equivalence relation on Z. List the equivalence classes for this equivalence relation.

6. Show that the relation "if and only if" on a nonempty collection of propositions is an equivalence relation.

7. List all equivalence relations on $S = \{1,2,3,4\}$.

2.3 MAPPINGS

Another type of relation that plays a unifying role in mathematics is a mapping (function). Intuitively we often consider a mapping as a rule that associates elements of one set uniquely with elements of another set. More formally:

2.3.1 DEFINITION. Let A and B be nonempty sets. A *mapping* α from A to B is a relation α from A to B satisfying these conditions:
1. for each $a \in A$ there exists $b \in B$ such that $(a,b) \in \alpha$, and
2. $(a,b_1) \in \alpha$ and $(a,b_2) \in \alpha$ imply $b_1 = b_2$.

In short, $\alpha \subseteq A \times B$, in which each $a \in A$ is related to exactly one $b \in B$. Symbolically, we write $\alpha: A \rightarrow B$, read "α is a mapping from A to B" or, more simply, "α is a map from A to B." If $(a,b) \in \alpha$, we call b the *image* of a under α and write $b = a\alpha$ (traditionally written $b = \alpha(a)$). Additionally, a is called a *pre-image* of b (why not *the* pre-image of b?). The convention of using right-handed notation has an advantage that will become apparent later in this section. The sets A and B are called the *domain* of α, denoted dom α, and the *terminal set* of α respectively.

2.3.2 DEFINITION. Let $\alpha: A \rightarrow B$. The *range* of α, denoted by $A\alpha$, is the subset of B consisting of all second coordinates of the ordered pairs in α. Symbolically, $A\alpha = \{a\alpha \,|\, a \in A\}$.

If the elements of dom α all have the same image, then the range is clearly a singleton and α is called a *constant map*. The range of

α is not necessarily a singleton; it could conceivably be any subset of the terminal set. For $\alpha: A \to B$ and $S \subseteq A$, we define $S\alpha$ to be the collection of images of elements in S; that is, $S\alpha = \{s\alpha \mid s \in S\}$. Clearly $S\alpha \subseteq A\alpha$.

EXAMPLES

1. Let $A = \{1,2,3\}$, $B = \{2,4,6,8\}$, and $\alpha = \{(1,2),(2,4),(3,6)\}$. Clearly $\alpha: A \to B$ and $A\alpha = \{2,4,6\}$. For $S = \{1,2\}$, $S\alpha = \{2,4\}$.

2. Let A and B be as in Example 1 and let $\alpha = \{(1,2),(2,3),(2,4)\}$. α is not a map from A to B, since the image of 2 is not unique. Also, 3 has no image.

3. Let $A = B = F$. Define α as follows: For all $x \in F$, let $x\alpha = x^2$ (i.e., $\alpha = \{(x,x^2) \mid x \in F\}$). Then $\alpha: A \to B$ and $A\alpha$ is the set of non-negative real numbers.

4. Let S be a nonempty set. $i_S: S \to S$ defined by $xi_S = x$ for each $x \in S$ is the *identity map* on S ($i_S = \{(x,x) \mid x \in S\}$).

The presence of either or both of the following properties will frequently be essential in using maps to aid in the investigation of the mathematical systems that we will consider. Mappings will serve as one of the three main tools for studying such systems and will be emphasized throughout this book.

2.3.3 DEFINITION. Let $\alpha: A \to B$. α is an *onto map* if for each $b \in B$ there exists at least one $a \in A$ such that $a\alpha = b$. Equivalently, α is onto if the range of α is the entire terminal set ($A\alpha = B$).

2.3.4 DEFINITION. Let $\alpha: A \to B$. α is a *one-to-one map* if $a\alpha = b\alpha$ implies $a = b$ for $a,b \in A$. Equivalently, α is 1–1 (one-to-one) if each $a \in A$ maps to a different $b \in B$.

EXAMPLES

1. Let $\alpha: A \times B \to A$ defined by $(a,b)\alpha = a$ for all $(a,b) \in A \times B$. Clearly α is an onto map. However, $(a,b_1)\alpha = a = (a,b_2)\alpha$ does not imply $(a,b_1) = (a,b_2)$ unless B contains only one element; therefore, in general, α is not 1–1. This mapping is called the *projection* of $A \times B$ onto A.

2. Let $\alpha: Z \to Z$ defined by $x\alpha = 2x + 6$ for all $x \in Z$. α is not onto, since no odd numbers are images—that is, $Z\alpha = \{x \mid x$ is an even integer$\}$. α is 1–1, however, since $a\alpha = b\alpha$ implies $2a + 6 = 2b + 6$, which in turn implies $a = b$.

3. Let $S \neq \emptyset$. $i_S: S \to S$ is both 1–1 and onto ($i_S = \{(x,x) \mid x \in S\}$).

Mappings with both the 1–1 and onto properties, called 1–1 *correspondences*, lead in a rather natural way to new mappings of interest to us.

2.3.5 DEFINITION. Let $\alpha: A \to B$, which is 1–1 and onto. α^{-1} (read α inverse) is the mapping defined by $b\alpha^{-1} = a$ if and only if $a\alpha = b$. Symbolically, $\alpha^{-1} = \{(b,a) \mid (a,b) \in \alpha\}$. $\alpha^{-1}: B \to A$, which is also 1–1 and onto.

Intuitively the inverse map reverses the action (direction) of the map. The inverse of squaring a positive number is taking the positive square root, and the inverse of adding 3 is subtracting 3.

By defining a mapping as a set of ordered pairs, we inherit equality of maps. We have previously defined set equality; thus mappings α and β are equal if and only if they are composed of the same ordered pairs. Equivalently, maps α and β are equal if the following conditions hold:

1. dom α = dom β, and
2. For each $a \in$ dom α, $a\alpha = a\beta$.

This gives us an alternate method for proving maps equal: show that they have the same domains and the same action.

We now consider a natural way of combining two mappings to obtain a new mapping.

2.3.6 DEFINITION. Let $\alpha: A \to B$ and $\beta: B \to C$. The *composition* of α and β, denoted by $\alpha\beta$, is a mapping from A to C defined by $a(\alpha\beta) = (a\alpha)\beta$ for each $a \in A$.

Note that for maps α and β to be conformable for composition in that order, we must have $A\alpha \subseteq$ dom β (see 2.3.7). What relationship must exist between $B\beta$ and dom α if β and α are conformable for composition in this order? Many important applications of composition arise when $A = B = C$. Clearly, then, both $\alpha\beta$ and $\beta\alpha$ exist (but are not necessarily equal; see Exercise 2 of this section).

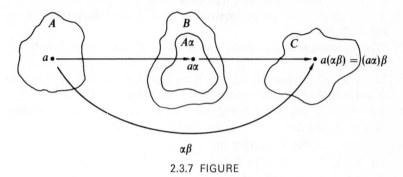

$\alpha\beta$

2.3.7 FIGURE

The advantage of right-hand notation is now apparent. Traditional notation forces the composition of α and β to appear as $\alpha\beta(a) = \beta(\alpha(a))$ for all $a \in A$. This suggests a reverse order which is definitely not present (see 2.3.7).

EXAMPLE. Let $A = \{1,2,3,4\}$, $B = \{2,4,6,8,10\}$, and $C = \{3,5,7,9,11\}$. Define $n\alpha = 2n$ for all $n \in A$ and $n\beta = n + 1$ for all $n \in B$. Then $\alpha: A \to B$, $\beta: B \to C$, and α and β are conformable for composition because $A\alpha \subseteq$ dom $\beta = B$. Now $\alpha\beta: A \to C$, and for each $n \in A, n(\alpha\beta) = (n\alpha)\beta = (2n)\beta = 2n + 1$. In particular, $1(\alpha\beta) = 3$, $2(\alpha\beta) = 5$, and so on. Are β and α conformable for composition? Why? Is the range of $\alpha\beta$ also the range of β; i.e., is $A\alpha\beta = B\beta$?

EXERCISES

1. Which of the following mappings are 1–1? Which are onto?
    (a) $\alpha: F \to F_+$ defined by $x\alpha = x^2 + 1$ for all $x \in F$.
    (b) $\beta: F \to F$ defined by $x\beta = x^3$ for all $x \in F$.
    (c) $\gamma: F \to F$ defined by $x\gamma = x - 1$ for all $x \in F$.
2. Using β and γ as in Exercise 1:
    (a) find $\beta\gamma$,
    (b) find $\gamma\beta$ (Is $\beta\gamma = \gamma\beta$?),
    (c) find γ^{-1}.

†3. Prove that the two ways of establishing equality of mappings as given in this section are equivalent.

4. Let $\alpha: A \to B$ and $\beta: B \to C$. Verify that $\alpha\beta$ is a mapping. Then prove:
   (a) if α is 1–1 and β is 1–1, then $\alpha\beta$ is 1–1;
   (b) if α is onto and β is onto, then $\alpha\beta$ is onto.
   This establishes that the composition of two 1–1 correspondences is a 1–1 correspondence.

5. Give an example of a mapping α such that $\alpha: Z \to Z_+$, which is both 1–1 and onto.

6. Let $\rho: S \to T$. Define the relation R on S as follows: For $x, y \in S$, xRy if and only if $x\rho = y\rho$. Prove that R is an equivalence relation on S.

†7. Let $\alpha: A \to B$, which is 1–1 and onto. Show that $\alpha\alpha^{-1} = i_A$, the identity map on A, and that $\alpha^{-1}\alpha = i_B$, the identity map on B.

8. Let $\alpha: A \to B$. Prove the following:
   (a) if $S \subseteq T \subseteq A$, then $S\alpha \subseteq T\alpha$;
   (b) if $S, T \subseteq A$, then $(S \cup T)\alpha = S\alpha \cup T\alpha$.

†9. Let $\beta: A \to B$ and $X, Y \subseteq A$. Show that $(X \cap Y)\beta \subseteq X\beta \cap Y\beta$. When does equality hold?

2.4   OPERATIONS

Just as we specialized relations to obtain mappings we now specialize mappings to obtain operations. The importance of operations cannot be overestimated. As we shall see in Chapter 3, their mere presence transforms an ordinary set into a mathematical system.

2.4.1 DEFINITION. An *operation* on a nonempty set A is a mapping from $A \times A$ to A.

We are actually defining a special type of operation called a binary operation on A. A ternary operation, for example, would be a mapping from $A \times A \times A$ to A. Operation in this book will be understood to mean binary operation.

The intuitive idea of an operation on A is the combining of arbitrary elements a and b of A in some prescribed way to obtain a unique element c of A. If $*$ is an operation on A, we will usually write $a * b = c$ rather than $(a,b) * = c$. This notational convenience is a generalization of that used for addition, subtraction, and multiplication in the familiar number systems.

EXAMPLES

1. Addition $(+)$ is an operation on Z since $+: Z \times Z \to Z$. We write $1 + 2 = 3$ rather than $(1,2)+ = 3$. Making "circular" use of the symbol $+$, we can write $+ = \{((a,b),c) \mid a,b,c \in Z \text{ and } a + b = c\}$.

2. Subtraction and multiplication are examples of operations on Z. Division, however, is not an operation on Z. Why not?

3. Let $P(S)$ be the *power set* of S; that is, $P(S) = \{T \mid T \subseteq S\}$. $\cup$ and $\cap$ are both operations on $P(S)$.

4. Let S be a nonempty set and $M(S)$ be the *set of all maps* on S; that is $M(S) = \{\alpha \mid \alpha: S \to S\}$. Composition is an operation on $M(S)$. To see this, note that $\alpha, \beta \in M(S)$ imply $\alpha\beta \in M(S)$.

A more general idea of an operation is a mapping from $A \times A$ to B. However, to adopt such a definition necessitates talking about closed operations (i.e., $B \subseteq A$). We built closure into our definition, since the need for operations that are not closed does not arise in this book.

Several properties that operations might possess are of interest to us.

2.4.2 DEFINITION. Let $*$ be an operation on A.
1. $*$ is *commutative* on A if for all $a,b \in A$, $a * b = b * a$.
2. $*$ is *associative* on A if for all $a,b,c \in A$ we have
$$(a * b) * c = a * (b * c).$$

If $*$ is associative on A, we often write $a * b * c$ for $(a * b) * c$ or $a * (b * c)$ since equality holds. For example, $1 + 2 + 3$ offers no confusion, since $(1 + 2) + 3 = 1 + (2 + 3) = 6$. However, $1 - 2 - 3$ is ambiguous, since $(1 - 2) - 3 = -4$ and $1 - (2 - 3) = 2$. Thus if $*$ is not associative, we cannot use $a * b * c$ without parentheses.

EXAMPLES

1. Addition is both commutative and associative on Z.
2. Subtraction is neither associative nor commutative on Z.
3. $\cup$ and $\cap$ are each commutative and associative on $P(S)$, the power set of S.

The next theorem and corollary are included to show that composition of mappings is an associative operation.

2.4.3 THEOREM. If α, β, and γ are mappings conformable for composition in that order, then $(\alpha\beta)\gamma = \alpha(\beta\gamma)$.

Proof: Let $\alpha: A \to B$, $\beta: B \to C$, and $\gamma: C \to D$. From the definition of composition of maps, we note that the domain of the composite map is the domain of the first map. Therefore dom $(\alpha\beta)\gamma$ = dom $\alpha\beta$ = dom α and dom $\alpha(\beta\gamma)$ = dom α. Hence the maps $(\alpha\beta)\gamma$ and $\alpha(\beta\gamma)$ have the same domain. Let $a \in A$. Then $a((\alpha\beta)\gamma) = (a(\alpha\beta))\gamma$ = $((a\alpha)\beta)\gamma = (a\alpha)\beta\gamma = a(\alpha(\beta\gamma))$, and the maps $(\alpha\beta)\gamma$ and $\alpha(\beta\gamma)$ have the same action and thus are equal.▲

2.4.4 COROLLARY. Composition is an associative operation on $M(A)$, the set of all mappings on the nonempty set A.

Proof: Exercise 4 of this section.▲

EXERCISES

1. Is division an operation on Q? On Q_0? If so, is division commutative on Q_0? Is it associative on Q_0?
2. Is subtraction an operation on Z_+? On the even integers?
†3. Show, via a counterexample, that composition on $M(S)$ is not a commutative operation.
4. Prove 2.4.4.
5. Distinguish between a mapping on a set A and an operation on A.
6. Let $*$ be an operation on A. If $*$ is associative and commutative and $a,b,c \in A$, show that $a * (b * c) = (c * a) * b$.

2.5   SPECIAL ELEMENTS

Once we have an operation on a set, we begin looking for elements that behave in a special way. For example, $0 \in Z$ behaves in a special way relative to addition on Z. The integer one behaves in the same way relative to multiplication on Z. Elements possessing this property warrant further attention.

2.5.1 DEFINITION. Let $*$ be an operation on A. $l \in A$ is a *left identity* for $*$ if $l * a = a$ for each $a \in A$. $r \in A$ is a *right identity* for $*$ if $a * r = a$ for each $a \in A$. If $e \in A$ is both a left and right identity for $*$, we call e an *identity* for $*$.

EXAMPLES

1. Consider $P(A)$, the power set of A. The empty set acts as an identity for the operation of union on $P(A)$. A itself acts as an identity for the operation of intersection on $P(A)$.

2. Consider $M(A)$, the set of all mappings on A. i_A, the identity map on A, is an identity for composition on $M(A)$.

3. Zero is a right identity for subtraction on Z and one is a right identity for division on Q_0.

4. Define $*$ on Z as follows: for all $a,b \in Z$

$$a * b = \begin{cases} a \text{ if } b \text{ is even.} \\ b \text{ if } b \text{ is odd.} \end{cases}$$

$*$ is an operation on Z and every even integer is a right identity. Does there exist a left identity for $*$? Is $*$ commutative on Z? Is $*$ associative on Z?

2.5.2 THEOREM. Let $*$ be an operation on A. If l is a left identity and r is a right identity for $*$, then $l = r$ is an identity for $*$.

Proof: Exercise 1 of this section.▲

2.5.3 COROLLARY. Let $*$ be an operation on A. If e is an identity for $*$, then e is unique.

Proof: Exercise 2 of this section.▲

An identity's special behavior is the same relative to every element in the set. The next type of element that we consider is only required to behave in its special way relative to one particular element in the set. The following definition identifies this special element which may exist only when an identity is present.

2.5.4 DEFINITION. Let $*$ be an operation on A with identity e. For $a,b,c \in A$, b is a *left inverse* of a if $b * a = e$ and c is a *right inverse* of a if $a * c = e$. If there exists $d \in A$ so that d is both a left and a right inverse of a, we say that d is an *inverse* of a and write $d = a^{-1}$ (the only exception being $d = -a$ when the operation is denoted by $+$).

EXAMPLES

1. Relative to addition on Z, the inverse of a is $-a$ for all $a \in Z$.

2. Relative to multiplication on Z, only 1 and -1 have inverses. What are they?

3. Relative to multiplication on Q, if $a \in Q$ and $a \neq 0$, then $a^{-1} = 1/a$.

4. If $*$ is an operation on A and e is the identity for $*$, then e is its own inverse.

The equality between left and right identities for the operation $*$ on A depended only on their existence. To establish equality between left and right inverses for a given element a (relative to $*$), we need both their existence and the associativity of $*$.

2.5.5 THEOREM. Let $*$ be an associative operation on A with identity e. If $a \in A$ has both a left and a right inverse, then they are equal.

Proof: Exercise 4 of this section.▲

2.5.6 COROLLARY. Let $*$ be an associative operation on A with identity e. If $a \in A$ has an inverse, then it is unique.

Proof: Exercise 5 of this section.▲

It is inconvenient to give examples of "one-sided inverses" at this time, since the familiar operations that have an identity are also associative and commutative. For this reason, we defer such examples to the next section.

EXERCISES

1. Prove 2.5.2.
2. Prove 2.5.3.

†3. Let $*$ be a commutative operation on A with right identities r_1 and r_2. Prove that $r_1 = r_2$.

†4. Prove 2.5.5.

5. Prove 2.5.6.

†6. Define $*$ on Z as follows: For $a,b \in Z$, $a * b = a + b - ab$.
    (a) Is $*$ commutative on Z?
    (b) Is $*$ associative on Z?
    (c) Does there exist an identity relative to $*$?
    (d) Does $a \in Z$ have an inverse relative to $*$?

2.6 OPERATION TABLES

Suppose that A is a "small" finite set and $*$ is an operation on A. We can write out a complete description of $*$ as follows.

Let $A = \{a,b,c\}$ and $*$ be such that

$$a * a = a \qquad b * a = b \qquad b * c = a$$
$$a * b = b \qquad c * a = c \qquad c * b = a$$
$$a * c = c \qquad b * b = c \qquad c * c = b.$$

Obviously, even with a set of three elements this method of displaying information is rather awkward. However, the situation is easily improved by recording the information in the form of a table.

$*$	a	b	c
a	a	b	c
b	b	c	a
c	c	a	b

In addition, the information is easily retrieved from the table; that is, for $b * c$ simply locate the intersection of row b with column c and find a ($b * c = a$).

It is relatively easy to determine the existence of special elements from a properly constructed table and also whether or not the operation is commutative. The operation $*$ as defined in this section is commutative, whereas $\boxdot$ (read "square dot") defined by the following table is not commutative.

$\boxdot$	a	b	c
a	b	c	b
b	a	b	a
c	c	a	c

Commutativity of the operation corresponds to symmetry in the table with respect to the major diagonal (upper-left to lower-right corner); that is, if we fold the table along the major diagonal, then the operation is commutative only when the corresponding entries are the same.

The existence of a left identity can be determined simply by locating a row that is a repetition of the column headings. The element a is a left identity for $*$ above, while there is no left identity for $\boxdot$ on A. A right identity is found (if one exists) by locating a column that is a repeat of the row headings. Again, a is a right identity for $*$, while $\boxdot$ has no right identity. Putting these two ideas together yields a method for determining the existence (or nonexistence) of an identity relative to a given operation.

Once an identity has been determined, inverses can easily be located if they exist. To find a right inverse for a given element, search across the row headed by this element until the identity is located (the identity will be in this row at least once if a right inverse exists). Whatever appears as that particular column heading is a desired right inverse. For a left inverse, interchange row with column and column with row in the preceding.

EXAMPLE. Define $\odot$ (read "circle dot") on $A = \{a,b,c\}$ as follows:

$\odot$	a	b	c
a	a	b	c
b	b	a	a
c	c	b	c

The element a is an identity and is its own inverse. The element b has two right inverses, b and c, and one left inverse, b. Finally, c has one left inverse, b, and no right inverse.

EXERCISES

1. If possible, use an operation table to define $\odot$ on $A = \{a,b,c\}$ such that
   (a) a, b, and c are each left identities,
   (b) a is a left identity and b is a right identity,
   (c) a is the identity and each element is its own inverse,
   (d) $\odot$ is associative, a is the identity, and each element is its own inverse.

†2. Use an operation table to define $\boxdot$ on $S = \{e,a,b,c\}$ such that $\boxdot$ is commutative, e is the identity, and each element is its own inverse.

†3. For the operation $*$ on A, given in this section, what information can you glean from the operation table?

2.7 CONGRUENT INTEGERS

Familiarity with the basic algebraic properties of the integers, the rationals, and the reals is assumed. The following condensation of these properties is included as a short review. Addition and multiplication are defined on Z, Q, and F, and

1. addition and multiplication are commutative,
2. addition and multiplication are associative,
3. addition and multiplication have 0 and 1 as unique identities respectively,
4. each element has a unique additive inverse, and
5. multiplication is distributive over addition.

In addition to the common properties, unique multiplicative inverses exist for each nonzero element in Q and F.

The well ordering of Z_+ is another basic property that we will use from time to time. Well ordering refers to the intuitively obvious fact that every nonempty subset of Z_+ has a smallest element with respect to the relation "is less than."

Throughout this book we will refer to the familiar number systems for examples of many theoretical concepts. The remainder of this section is included to provide additional examples and, hopefully, to help ferment the ideas of equivalence relation and equivalence class.

The development of congruent integers that follows will make use of the division algorithm as well as the basic properties of Z.

2.7.1 THE DIVISION ALGORITHM. For given integers a and b with $b > 0$, there exist unique integers q and r such that $a = bq + r$ and $0 \leq r < b$.

Proof: Let $S = \{a - bx \mid a - bx \geq 0\}$. Since $b \in Z_+$, $b \geq 1$; thus $b|a| \geq |a| \geq a$ and $a + b|a| \geq 0$. Taking $x = -|a|$, we have $a - bx = a + b|a| \geq 0$. Therefore $(a + b|a|) \in S$ so that $S \neq \emptyset$. If

$0 \in S$, 0 is the smallest element of S; otherwise $S \subseteq Z_+$ and has a smallest element since Z_+ is well ordered. In either case, S has a smallest element, say r. Let q be such that $r = a - bq$. We now have $a = bq + r$ with $0 \le r$. To show that $r < b$, assume by way of contradiction that $r \ge b$. Then $a - b(q + 1) = (a - bq) - b = r - b \ge 0$ and $(r - b) \in S$, which contradicts the fact that r is the smallest element of S. Therefore $r < b$.

For uniqueness of q and r, suppose that q' and r' are such that $a = bq' + r'$ with $0 \le r' < b$. Then $bq + r = bq' + r'$ and $b(q - q') = r' - r$. Furthermore, $b|q - q'| = |r' - r| < b$, since $0 \le r < b$ and $0 \le r' < b$. This requires $|q - q'| = 0$; thus $q = q'$ and $r = r'$.▲

When a is the dividend and b the divisor, the integers q and r are called the quotient and the remainder respectively. Consider the integers 17 and 5. We can write $17 = 5 \cdot 3 + 2$. Also, $17 = 5 \cdot 2 + 7$. Does this contradict the uniqueness which 2.7.1 guarantees?

2.7.2 DEFINITION. Let $m \in Z_+$. Integers a and b are *congruent modulo m* if $a - b$ is divisible by m; that is, if there exists $k \in Z$ such that $a - b = km$. Symbolically, we write $a \underset{m}{\sim} b$.

EXAMPLE. $5 \underset{3}{\sim} 11$ since 3 divides -6; that is, since $5 - 11 = -6 = (-2)(3)$. Also, $17 \underset{12}{\sim} 5$ and $387 \underset{73}{\sim} 241$.

2.7.3 THEOREM. Congruence modulo m is an equivalence relation on Z.

Proof: The reflexive and symmetric properties of $\underset{m}{\sim}$ follow immediately from 2.7.2 (see Exercise 2 of this section). For transitivity, let $x, y, z \in Z$ with $x \underset{m}{\sim} y$ and $y \underset{m}{\sim} z$. Then there exist integers k_1 and k_2 such that $x - y = k_1 m$ and $y - z = k_2 m$. Adding these equations, we have $x - z = (k_1 + k_2)m$; thus $x \underset{m}{\sim} z$.▲

Frequently equivalence relations fail to partition infinite sets into a finite number of equivalence classes. For the equivalence relation $\underset{m}{\sim}$, however, we have the following theorem.

2.7.4 THEOREM. Z_m, the partition associated with $\underset{m}{\sim}$, contains exactly m cells.

Proof: We will show that $C = \{\bar{0}, \bar{1}, \ldots, \overline{m-1}\}$ contains m distinct cells and $C = Z_m$. For $0 \leq i < m$, $0 \leq j < m$ and $i \neq j$, we have $i \not\underset{m}{\sim} j$; therefore $\bar{i} \neq \bar{j}$ and $\bar{0}, \bar{1}, \ldots, \overline{m-1}$ are distinct cells. Clearly $C \subseteq Z_m$. Thus it remains to be shown that $Z_m \subseteq C$. Let $\bar{a} \in Z_m$. For integers a and $m > 0$, there exist $q, r \in Z$ such that $a = mq + r$ and $0 \leq r < m$ by 2.7.1 Now $a - r = mq$; so $a \underset{m}{\sim} r$ and $\bar{a} = \bar{r}$ by 2.7.3. Finally, $\bar{a} = \bar{r} \in C$ since $0 \leq r \leq m - 1$. Therefore $C = Z_m$. ▲

EXAMPLES

1. The equivalence classes of $\underset{4}{\sim}$ are $\bar{0} = \{\ldots, -8, -4, 0, 4, 8, \ldots\}$, $\bar{1} = \{\ldots, -7, -3, 1, 5, 9, \ldots\}$, $\bar{2}$, and $\bar{3}$. Note that $\bar{4} = \bar{0}$, $\bar{5} = \bar{1}$, etc.

2. The associated partition for $\underset{2}{\sim}$ on Z is $Z_2 = \{\bar{0}, \bar{1}\}$, where $\bar{0} = \{\ldots, -4, -2, 0, 2, 4, \ldots\}$ and $\bar{1} = \{\ldots, -3, -1, 1, 3, \ldots\}$.

2.7.5 **LEMMA.** If $a \underset{m}{\sim} b$, then $a + x \underset{m}{\sim} b + x$ and $ax \underset{m}{\sim} bx$ for all $x \in Z$.

Proof: Exercise 3 of this section. ▲

2.7.6 **LEMMA.** If $a \underset{m}{\sim} b$ and $c \underset{m}{\sim} d$, then $a + c \underset{m}{\sim} b + d$ and $ac \underset{m}{\sim} bd$.

Proof: $a \underset{m}{\sim} b$ and $c \underset{m}{\sim} d$; so, by 2.7.5, $a + c \underset{m}{\sim} b + c$ and $b + c \underset{m}{\sim} b + d$. $\underset{m}{\sim}$ is transitive; therefore $a + c \underset{m}{\sim} b + d$. Similarly, $ac \underset{m}{\sim} bd$. ▲

These lemmas enable us to define and investigate the following operations on Z_m, the partition associated with $\underset{m}{\sim}$.

2.7.7 **DEFINITION.** Let $\bar{a}, \bar{b} \in Z_m$. We define *addition modulo* m, $+_m$, and *multiplication modulo* m, $\cdot_m$, as follows:

$$\bar{a} +_m \bar{b} = \overline{a + b} \quad \text{and} \quad \bar{a} \cdot_m \bar{b} = \overline{ab}.$$

2.7.8 **THEOREM.** Addition modulo m and multiplication modulo m are operations on Z_m.

Proof: We establish the result for $+_m$ by showing uniqueness of the image of $(\bar{a}, \bar{b})$. Let $\bar{a} = \bar{x}$ and $\bar{b} = \bar{y}$. Then $a \underset{m}{\sim} x$ and $b \underset{m}{\sim} y$. By

2.7.6, $a + b \underset{m}{\sim} x + y$ whence $\overline{a + b} = \overline{x + y}$. The argument is similar for $\cdot_m$. ▲

This theorem establishes that $+_m$ is an "adding" of classes and not just representatives as it may appear. In fact, "adding classes" is independent of particular class representatives.

Similarly, $\cdot_m$ is truly a "multiplication" of classes rather than simply a multiplication of class representatives.

Some basic properties of the operations $+_m$ and $\cdot_m$ on Z_m are given in the next two theorems.

2.7.9 THEOREM. Addition modulo m is associative and commutative on Z_m. Furthermore, $\overline{0}$ is the identity for $+_m$ and each cell in Z_m has an inverse relative to $+_m$.

Proof: Exercise 4 of this section. ▲

2.7.10 THEOREM. Multiplication modulo m is associative and commutative on Z_m. Furthermore, $\overline{1}$ is the identity for $\cdot_m$.

Proof: Exercise 5 of this section. ▲

EXAMPLE. Consider Z_6. The six equivalence classes are

$$\overline{0} = \{0, \pm 6, \pm 12, \ldots\},$$
$$\overline{1} = \{1, 1 \pm 6, 1 \pm 12, \ldots\},$$
$$\overline{2} = \{2, 2 \pm 6, 2 \pm 12, \ldots\}, \overline{3}, \overline{4}, \text{ and } \overline{5}.$$

The operations $+_6$ and $\cdot_6$ are displayed in 2.7.11.

2.7.11 TABLES

$+_6$	$\overline{0}$	$\overline{1}$	$\overline{2}$	$\overline{3}$	$\overline{4}$	$\overline{5}$
$\overline{0}$	$\overline{0}$	$\overline{1}$	$\overline{2}$	$\overline{3}$	$\overline{4}$	$\overline{5}$
$\overline{1}$	$\overline{1}$	$\overline{2}$	$\overline{3}$	$\overline{4}$	$\overline{5}$	$\overline{0}$
$\overline{2}$	$\overline{2}$	$\overline{3}$	$\overline{4}$	$\overline{5}$	$\overline{0}$	$\overline{1}$
$\overline{3}$	$\overline{3}$	$\overline{4}$	$\overline{5}$	$\overline{0}$	$\overline{1}$	$\overline{2}$
$\overline{4}$	$\overline{4}$	$\overline{5}$	$\overline{0}$	$\overline{1}$	$\overline{2}$	$\overline{3}$
$\overline{5}$	$\overline{5}$	$\overline{0}$	$\overline{1}$	$\overline{2}$	$\overline{3}$	$\overline{4}$

$\cdot_6$	$\overline{0}$	$\overline{1}$	$\overline{2}$	$\overline{3}$	$\overline{4}$	$\overline{5}$
$\overline{0}$	$\overline{0}$	$\overline{0}$	$\overline{0}$	$\overline{0}$	$\overline{0}$	$\overline{0}$
$\overline{1}$	$\overline{0}$	$\overline{1}$	$\overline{2}$	$\overline{3}$	$\overline{4}$	$\overline{5}$
$\overline{2}$	$\overline{0}$	$\overline{2}$	$\overline{4}$	$\overline{0}$	$\overline{2}$	$\overline{4}$
$\overline{3}$	$\overline{0}$	$\overline{3}$	$\overline{0}$	$\overline{3}$	$\overline{0}$	$\overline{3}$
$\overline{4}$	$\overline{0}$	$\overline{4}$	$\overline{2}$	$\overline{0}$	$\overline{4}$	$\overline{2}$
$\overline{5}$	$\overline{0}$	$\overline{5}$	$\overline{4}$	$\overline{3}$	$\overline{2}$	$\overline{1}$

EXERCISES

†1. Find unique integers q and r so that $a = bq + r$ with $0 \leq r < b$ when
    (a) $a < b$ and $a \geq 0$,
    (b) $a = b$ and $b > 0$.

2. Show that $\tilde{m}$ is reflexive and symmetric on Z.

†3. Prove 2.7.5.

4. Prove 2.7.9.

5. Prove 2.7.10.

†6. What classes in Z_6 have inverses relative to $\cdot_6$ (see 2.7.11)?

7. Construct an operation table for $\cdot_7$. Which classes have inverses relative to $\cdot_7$?

8. Investigate the existence of inverses relative to $\cdot_m$ for $m = 3, 5, 7, 9$, and 11. Based on this investigation, make a conjective concerning the values of m for which each nonzero element in Z_m has a multiplicative inverse.

9. Show that $\cdot_m$ distributes over $+_m$ on Z_m; that is, show that
    $\bar{a} \cdot_m (\bar{b} +_m \bar{c}) = (\bar{a} \cdot_m \bar{b}) +_m (\bar{a} \cdot_m \bar{c})$ for all $\bar{a}, \bar{b}, \bar{c} \in Z_m$.

EXERCISES

1.

2.

3.

4.

5.

6.

7.

8.

CHAPTER 3

GROUPS

We are now ready to study our first algebraic system. An *algebraic system* consists of a nonempty set, one or more operations, one or more relations (with equality always included), and some particular basic properties that these operations and/or relations possess. Since a group is one of the most basic algebraic systems, we have chosen it as our first system for detailed study.

3.1· DEFINITION OF A GROUP

With only one operation, one relation, and three assumptions, one would not expect the group concept to hold the esteemed position in mathematics that it does. However, group theory is of fundamental importance to all branches of modern mathematics and has many sophisticated applications in other scientific areas.

3.1.1 DEFINITION. A *group* is an algebraic system consisting of a nonempty set G and an operation $*$ on G having the following properties:

1. $*$ is associative in G; that is, for all $a,b,c \in G$ we have $(a*b)*c = a*(b*c)$.

2. G has an identity relative to $*$; that is, there exists $e \in G$ such that $a*e = e*a = a$ for each $a \in G$.

3. Each element of G has an inverse; that is, for each $g \in G$ there exists $g^{-1} \in G$ such that $g*g^{-1} = g^{-1}*g = e$.

If in addition to (1), (2), and (3), $*$ is commutative in G—that is, $a*b = b*a$ for all $a,b \in G$—we say that the group is *commutative* or *abelian* (after the Norwegian mathematician Niels H. Abel, 1802–1829). Symbolically, we write $\langle G,* \rangle$ for a group. Technically, we should write $\langle G,*,= \rangle$; however, it is understood that equality is present in every algebraic system.

EXAMPLES OF GROUPS

1. From the familiar number systems we have $\langle Z,+ \rangle$, $\langle Q,+ \rangle$, $\langle Q_+,\cdot \rangle$, $\langle Q_0,\cdot \rangle$, $\langle F,+ \rangle$, $\langle F_0,\cdot \rangle$, and $\langle F_+,\cdot \rangle$.

2. The collection of all integral powers of a fixed nonzero real number t with usual multiplication $(t^x \cdot t^y = t^{x+y})$; that is, $\langle S,\cdot \rangle$ where $S = \{t^n \mid n \in Z$ and t is a fixed nonzero real number$\}$.

3. $\langle Z_m,+_m \rangle$ for each $m \in Z_+$.

4. $\langle \{0\},+ \rangle$, $\langle \{0\},\cdot \rangle$, $\langle \{1\},\cdot \rangle$.

5. The four fourth roots of unity, together with complex multiplication; that is, $\langle \{1,-1,i,-i\},\cdot \rangle$, where $i^2 = -1$.

6. The set of all nonsingular 2×2 matrices with matrix multiplication; that is,

$$\left\{ \begin{pmatrix} a & b \\ c & d \end{pmatrix} \middle| a,b,c,d \in F \text{ and } ad - bc \neq 0 \right\}$$

with multiplication defined by

$$\begin{pmatrix} s & t \\ u & v \end{pmatrix} \begin{pmatrix} w & x \\ y & z \end{pmatrix} = \begin{pmatrix} sw + ty & sx + tz \\ uw + vy & ux + vz \end{pmatrix}.$$

To show that this set, together with matrix multiplication, forms a group, let $\begin{pmatrix} a & b \\ c & d \end{pmatrix}$ and $\begin{pmatrix} e & f \\ g & h \end{pmatrix}$ be 2 × 2 matrices such that $ad - bc \neq 0$ and $eh - fg \neq 0$. Then

$$\begin{pmatrix} a & b \\ c & d \end{pmatrix}\begin{pmatrix} e & f \\ g & h \end{pmatrix} = \begin{pmatrix} ae + bg & af + bh \\ ce + dg & cf + dh \end{pmatrix}$$

and

$$(ae + bg)(cf + dh) - (af + bh)(ce + dg)$$
$$= aecf + aedh + bgcf + bgdh - afce - afdg - bhce - bhdg$$
$$= (aedh - bhce) + (bgcf - afdg)$$
$$= eh(ad - bc) + fg(ad - bc)$$
$$= (eh - fg)(ad - bc) \neq 0.$$

Therefore $\begin{pmatrix} a & b \\ c & d \end{pmatrix}\begin{pmatrix} e & f \\ g & h \end{pmatrix}$ is a nonsingular matrix and multiplication is an operation on this set. As for an identity,

$$\begin{pmatrix} 1 & 0 \\ 0 & 1 \end{pmatrix}\begin{pmatrix} a & b \\ c & d \end{pmatrix} = \begin{pmatrix} a & b \\ c & d \end{pmatrix}\begin{pmatrix} 1 & 0 \\ 0 & 1 \end{pmatrix} = \begin{pmatrix} a & b \\ c & d \end{pmatrix},$$

so $\begin{pmatrix} 1 & 0 \\ 0 & 1 \end{pmatrix}$ is the identity. Furthermore,

$$\begin{pmatrix} a & b \\ c & d \end{pmatrix}\begin{pmatrix} \dfrac{d}{ad - bc} & \dfrac{-b}{ad - bc} \\ \dfrac{-c}{ad - bc} & \dfrac{a}{ad - bc} \end{pmatrix} = \begin{pmatrix} 1 & 0 \\ 0 & 1 \end{pmatrix}$$

and

$$\begin{pmatrix} \dfrac{d}{ad - bc} & \dfrac{-b}{ad - bc} \\ \dfrac{-c}{ad - bc} & \dfrac{a}{ad - bc} \end{pmatrix}\begin{pmatrix} a & b \\ c & d \end{pmatrix} = \begin{pmatrix} 1 & 0 \\ 0 & 1 \end{pmatrix}.$$

Therefore, each matrix $\begin{pmatrix} a & b \\ c & d \end{pmatrix}$ with $ad - bc \neq 0$ has an inverse. To see that the operation is associative, let $\begin{pmatrix} s & t \\ u & v \end{pmatrix}$ and $\begin{pmatrix} w & x \\ y & z \end{pmatrix}$ be matrices with $sv - ut$ and $wz - xy$ different from zero. Then

$$\begin{pmatrix} a & b \\ c & d \end{pmatrix}\left[\begin{pmatrix} s & t \\ u & v \end{pmatrix}\begin{pmatrix} w & x \\ y & z \end{pmatrix}\right] = \begin{pmatrix} a & b \\ c & d \end{pmatrix}\begin{pmatrix} sw + ty & sx + tz \\ uw + vy & ux + vz \end{pmatrix}$$
$$= \begin{pmatrix} asw + aty + buw + bvy & asx + atz + bux + bvz \\ csw + cty + duw + dvy & csx + ctz + dux + dvz \end{pmatrix}$$

$$= \begin{pmatrix} (as + bu)w + (at + bv)y & (as + bu)x + (at + bv)z \\ (cs + du)w + (ct + dv)y & (cs + du)x + (ct + dv)z \end{pmatrix}$$

$$= \begin{pmatrix} as + bu & at + bv \\ cs + du & ct + dv \end{pmatrix} \begin{pmatrix} wx \\ yz \end{pmatrix}$$

$$= \left[\begin{pmatrix} a & b \\ c & d \end{pmatrix} \begin{pmatrix} s & t \\ u & v \end{pmatrix} \right] \begin{pmatrix} w & x \\ y & z \end{pmatrix}.$$

Examples 1 through 5 are commutative. Example 6, however, yields a noncommutative group since

$$\begin{pmatrix} 1 & 0 \\ 1 & 1 \end{pmatrix} \begin{pmatrix} 2 & 1 \\ 0 & 1 \end{pmatrix} = \begin{pmatrix} 2 & 1 \\ 2 & 2 \end{pmatrix}$$

while

$$\begin{pmatrix} 2 & 1 \\ 0 & 1 \end{pmatrix} \begin{pmatrix} 1 & 0 \\ 1 & 1 \end{pmatrix} = \begin{pmatrix} 3 & 1 \\ 1 & 1 \end{pmatrix}.$$

Axioms (2) and (3) in the definition of a group do not specify uniqueness. Yet the identity in a group is unique by 2.5.3 and the inverse of each element is also unique by 2.5.6. The remainder of this section is devoted to establishing some elementary group properties.

3.1.2 THEOREM. Let e be the identity in $\langle G, * \rangle$.
1. If $i \in G$ is a left (or right) identity for $*$, then $i = e$.
2. If $b \in G$ is a left (or right) inverse of $a \in G$, then $b = a^{-1}$.

Proof: 1. Let i be a left identity. Then $i * a = a$ for all $a \in G$. In particular, $i * e = e$. Since e is the identity in G, $i * e = i$. Hence $i = e$.

2. Let b be a left inverse of a. Then $b * a = e$. By multiplying on the right by a^{-1} (how do we know a^{-1} exists?), we have $(b * a) * a^{-1} = e * a^{-1}$. Since $*$ is associative, we have $b * (a * a^{-1}) = b * e$. Finally, $b * e = e * a^{-1}$ implies $b = a^{-1}$ since e is the identity in G. ▲

The next two results provide us with some of the "power" available in a group.

3.1.3 THEOREM. The left and right cancellation laws hold in $\langle G, * \rangle$—that is, for $a, b, c \in G$ with $a * b = a * c$ (or $b * a = c * a$), we have $b = c$.

Proof: Suppose that $a * b = a * c$. Multiplying by a^{-1} on the left yields $a^{-1} * (a * b) = a^{-1} * (a * c)$. Via associativity, $(a^{-1} * a) * b = (a^{-1} * a) * c$, which reduces to $b = c$. ▲

3.1.4 THEOREM. Linear equations have unique solutions in $\langle G, * \rangle$—that is, for $a, b \in G$, the equations $a * x = b$ and $y * a = b$ have the unique solutions $x = a^{-1} * b$ and $y = b * a^{-1}$.

Proof: It is easily verified that x and y as given in the theorem satisfy the equations. Uniqueness follows from the cancellation laws. ▲

3.1.5 COROLLARY. In $\langle G, * \rangle$, the inverse of a "product" is the "product" of the inverses in the reverse order—that is, for all $a, b \in G$ we have $(a * b)^{-1} = b^{-1} * a^{-1}$.

Proof: Exercise 5 of this section. ▲

Of course, if the group is abelian, then $(a * b)^{-1} = a^{-1} * b^{-1}$. For example, in $\langle Q_0, \cdot \rangle$ we have $(2 \cdot 3)^{-1} = 6^{-1} = \frac{1}{6} = \frac{1}{2} \cdot \frac{1}{3} = 2^{-1} \cdot 3^{-1}$.

3.1.6 COROLLARY. In $\langle G, * \rangle$, the inverse of the inverse of an element in G is the element—that is, for $g \in G$, $(g^{-1})^{-1} = g$.

Proof: Exercise 6 of this section. ▲

EXERCISES

1. Verify that the structures given in Examples 2, 3, and 5 form groups.
2. Let $\alpha, \beta \in M(F_0)$ be defined as follows: $x\alpha = 1/x$ and $x\beta = x$ for all $x \in F_0$. Show that $\{\alpha, \beta\}$, together with composition of maps, is a group.
†3. Which of the following structures are groups?
   (a) The set of even integers with addition.
   (b) The set of positive irrational numbers under multiplication.
   (c) $\{3m \mid m \in Z\}$ with the operation of addition.
   (d) Z with $\odot$ defined by $a \odot b = a + b + 1$ for all $a, b \in Z$.
   (e) Z with $\odot$ defined by $a \odot b = a + b - ab$ for all $a, b \in Z$.
4. Supply the details for the proof of 3.1.4.

5. Prove 3.1.5.

6. Prove 3.1.6.

7. Let e be the identity in $\langle G, * \rangle$. Prove that the group is abelian if $a * a = e$ for all $a \in G$; that is, each element of G is its own inverse. (*Hint:* Show that $b * a$ is the inverse of $a * b$.)

8. Prove that $G = Z \times Z$ with $\odot$ defined by $(a,b) \odot (c,d) = (a + c - 1, b + d)$ for $(a,b),(c,d) \in G$ is a commutative group.

†9. Show that $G = Z \times Z$ with $\odot$ defined by $(a,b) \odot (c,d) = (a + d, b + d)$ for all $(a,b),(c,d) \in G$ is not a group by exhibiting a right identity that is not an identity (see 3.1.2).

10. Let e be the identity in $\langle G, * \rangle$. An element $a \in G$ is *idempotent* if $a * a = a$. Prove that the only idempotent element in G is e.

†11. Show that $P(S)$, the power set of S, together with $+$ defined by $A + B = (A \cup B) \setminus (A \cap B)$ for all $A,B \in P(S)$, forms an abelian group.

12. In a group, all linear equations are solvable by 3.1.4. Give an example to show that the quadratic equation $a * x^2 = b$ (i.e., $a * x * x = b$) is not always solvable in a group. (*Hint:* Consider $\langle Q_0, \cdot \rangle$.)

†13. Let $*$ be an associative operation on a nonempty set S and suppose that the equations $a * x = b$ and $y * a = b$ have unique solutions for all $a,b \in S$. Show that S, together with $*$, is a group.

3.2   HOMOMORPHISM AND ISOMORPHISM

Three basic techniques will be employed to obtain information about groups.

1. A study of mappings between groups that preserve their respective group operations.

2. A study of some particular subsets of a group.

3. A study of group decompositions via equivalence relations.

This section initiates the first of these studies.

Intuitively, a mapping between $\langle G_1, * \rangle$ and $\langle G_2, \odot \rangle$ preserves the respective operations if the image of each "product" in G_1 is equal to the "product" of the corresponding images in G_2. More formally:

3.2.1 **DEFINITION.** The mapping $\alpha: G_1 \to G_2$ is a *homomorphism* (group homomorphism) from $\langle G_1, * \rangle$ to $\langle G_2, \odot \rangle$ if, for all $a,b \in G_1$, we have $(a * b)\alpha = (a\alpha) \odot (b\alpha)$.

$G_1\alpha$, the range of α, is called a *homomorphic image* of G_1; and if α is onto $(G_1\alpha = G_2)$, then $\langle G_1,* \rangle$ is said to be *homomorphic* onto $\langle G_2,\odot \rangle$. Schematically, for a homomorphism α, see Figure 3.2.2.

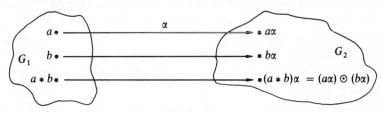

3.2.2 FIGURE

The following theorem verifies two basic properties of a homomorphism. In addition, it shows that not every mapping from one group to another possesses the homomorphism property.

3.2.3 THEOREM. If α is a homomorphism from $\langle G_1,* \rangle$ to $\langle G_2,\odot \rangle$, then

1. the image of the identity in G_1 is the identity in G_2, and
2. the image of g^{-1} in G_1 is $(g\alpha)^{-1}$ in G_2; that is, $(g^{-1})\alpha = (g\alpha)^{-1}$.

Proof: 1. Let e be the identity in G_1. Then $e = e * e$ and we have $e\alpha = (e * e)\alpha = (e\alpha) \odot (e\alpha)$. This says that $e\alpha \in G_2$ is idempotent. By Exercise 10 of 3.1, $e\alpha$ is the identity in G_2.

2. Let $g \in G_1$. Now $e\alpha = (g * g^{-1})\alpha = (g\alpha) \odot (g^{-1}\alpha)$. By 3.1.2.2, $g^{-1}\alpha = (g\alpha)^{-1}$.▲

The next theorem shows that we could adopt the convenience of considering only onto homomorphisms without any loss of generality.

3.2.4 THEOREM. If α is a homomorphism from $\langle G_1,* \rangle$ to $\langle G_2,\odot \rangle$, then $G_1\alpha$, the homomorphic image of G_1, together with $\odot$, is a group.

Proof: By 3.2.3, $G_1\alpha$ is nonempty, since the identity in G_2 is in $G_1\alpha$. Let $a',b' \in G_1\alpha$. Then there exist $a,b \in G_1$ such that $a' = a\alpha$ and $b' = b\alpha$ (why?). Now $a' \odot b' = (a\alpha) \odot (b\alpha) = (a * b)\alpha \in G_1\alpha$ and $\odot$ is an operation on $G_1\alpha$. Exercise 1 of this section establishes the existence of inverses. Finally, associativity of $\odot$ is inherited from G_2 and thus $G_1\alpha$, together with $\odot$, is a group.▲

EXAMPLES

1. Let $S = \{t^n \mid n \in Z$ and t is a fixed nonzero real number$\}$. $\eta: Z \to S$, defined by $a\eta = t^a$ for all $a \in Z$, is a homomorphism from $\langle Z, + \rangle$ onto $\langle S, \cdot \rangle$. To see this, let $a, b \in Z$. Then $(a + b)\eta = t^{a + b} = t^a \cdot t^b = (a\eta) \cdot (b\eta)$.

2. Define $\beta: Z \to Z$ by $a\beta = 2a$ for all $a \in Z$. This map is a homomorphism from $\langle Z, + \rangle$ to $\langle Z, + \rangle$. What group is the homomorphic image of Z under β?

3. The mapping $\alpha: Z \to Z$ defined by $a\alpha = 0$ for all $a \in Z$ is a homomorphism from $\langle Z, + \rangle$ to $\langle Z, + \rangle$.

4. The mapping $\eta: Z_8 \to Z_4$ defined by the following is a homomorphism from $\langle Z_8, +_8 \rangle$ onto $\langle Z_4, +_4 \rangle$. For each $\bar{x} \in Z_8$, $\bar{x}\eta = \bar{r}$, where $r \overset{\sim}{4} x$ and $\bar{r} \in Z_4$. In particular, $\bar{0}\eta = \bar{0}$, $\bar{1}\eta = \bar{1}$, $\bar{2}\eta = \bar{2}$, $\bar{3}\eta = \bar{3}$, $\bar{4}\eta = \bar{0}$, $\bar{5}\eta = \bar{1}$, $\bar{6}\eta = \bar{2}$, and $\bar{7}\eta = \bar{3}$. Clearly η is an onto map. As one instance of the homomorphism property, $(\bar{2} +_8 \bar{5})\eta = \bar{7}\eta = \bar{3}$ and $\bar{2}\eta +_4 \bar{5}\eta = \bar{2} +_4 \bar{1} = \bar{3}$ so $(\bar{2} +_8 \bar{5})\eta = \bar{2}\eta +_4 \bar{5}\eta$. In general, $(\bar{a} +_8 \bar{b})\eta = \bar{a}\eta +_4 \bar{b}\eta$ for all $\bar{a}, \bar{b} \in Z_8$.

We now consider the special case of a homomorphism α from $\langle G_1, * \rangle$ onto $\langle G_2, \odot \rangle$, which is a 1–1 correspondence. In this instance, the only "difference" between $\langle G_1, * \rangle$ and $\langle G_2, \odot \rangle$ will be the notation for the elements and the operations; that is, G_2 is nothing more than G_1 in disguise.

3.2.5 DEFINITION. A 1–1 homomorphism from $\langle G_1, * \rangle$ onto $\langle G_2, \odot \rangle$ is an *isomorphism* from $\langle G_1, * \rangle$ onto $\langle G_2, \odot \rangle$. In this case, these groups are said to be *isomorphic* and we write $\langle G_1, * \rangle \approx \langle G_2, \odot \rangle$.

Notice that when α is an isomorphism, α^{-1} exists, since α is 1–1 and onto. In addition, we have α^{-1} is an isomorphism (see Exercise 2 of this section).

EXAMPLE. Consider $\langle Z_4, +_4 \rangle$ and $\langle S, \cdot \rangle$, where $S = \{1, -1, i, -i\}$. The operation tables are given in 3.2.6. Define $\alpha: Z_4 \to S$ by $\bar{0}\alpha = 1$, $\bar{1}\alpha = i$, $\bar{2}\alpha = -1$, and $\bar{3}\alpha = -i$. Clearly the identities and inverses correspond via α, and α is both 1–1 and onto; so we might suspect that α is an isomorphism. To actually verify the homomorphism

3.2.6 TABLE

$+_4$	$\bar{0}$	$\bar{1}$	$\bar{2}$	$\bar{3}$
$\bar{0}$	$\bar{0}$	$\bar{1}$	$\bar{2}$	$\bar{3}$
$\bar{1}$	$\bar{1}$	$\bar{2}$	$\bar{3}$	$\bar{0}$
$\bar{2}$	$\bar{2}$	$\bar{3}$	$\bar{0}$	$\bar{1}$
$\bar{3}$	$\bar{3}$	$\bar{0}$	$\bar{1}$	$\bar{2}$

$\cdot$	1	i	-1	$-i$
1	1	i	-1	$-i$
i	i	-1	$-i$	1
-1	-1	$-i$	1	i
$-i$	$-i$	1	i	-1

property for α at this time would require individual consideration for each pair of elements in Z_4. Since there are sixteen such pairs, doing so would be tedious at best. For the present, we accept the fact that α is an isomorphism (verification is deferred to Exercise 9 in 3.7). We now see that "dressing" the elements of Z_4 and $+_4$ via the action of α gives us $\langle S, \cdot \rangle$. Thus $\langle S, \cdot \rangle$ is simply $\langle Z_4, +_4 \rangle$ in disguise.

3.2.7 THEOREM. The relation $\approx$ is an equivalence relation on a nonempty collection $\mathcal{G}$ of groups.

Proof: 1. Let $\langle G, * \rangle \in \mathcal{G}$. The identity map on G, i_G, is an isomorphism from $\langle G, * \rangle$ onto itself; thus $\approx$ is reflexive on $\mathcal{G}$.

2. See Exercise 2 of this section for symmetry of $\approx$ on $\mathcal{G}$.

3. For transitivity, let $\langle G_1, * \rangle$, $\langle G_2, \odot \rangle$, $\langle G_3, \boxdot \rangle \in \mathcal{G}$ so that $\langle G_1, * \rangle \approx \langle G_2, \odot \rangle$ and $\langle G_2, \odot \rangle \approx \langle G_3, \boxdot \rangle$. Then there exist $\alpha: G_1 \rightarrow G_2$ and $\beta: G_2 \rightarrow G_3$ so that α and β are isomorphisms. We show that $\alpha\beta$ is an isomorphism from $\langle G_1, * \rangle$ onto $\langle G_3 \boxdot \rangle$. By Exercise 4 of 2.3, we have $\alpha\beta$ is 1–1 and onto. If $a, b \in G_1$, then

$$(a * b)\alpha\beta = [(a * b)\alpha]\beta$$
$$= [(a\alpha) \odot (b\alpha)]\beta \text{ (Why?)}$$
$$= [(a\alpha)\beta] \boxdot [(b\alpha)\beta] \text{ (Why?)}$$
$$= [a(\alpha\beta)] \boxdot [b(\alpha\beta)]. \blacktriangle$$

EXERCISES

1. For 3.2.4, verify the existence of inverses in $G_1\alpha$.
2. Prove that $\approx$ is symmetric on a nonempty collection $\mathcal{G}$ of groups.
†3. If $\langle G_2, \odot \rangle$ is a homomorphic image of $\langle G_1, * \rangle$ and $\langle G_1, * \rangle$ is abelian, prove that $\langle G_2, \odot \rangle$ is abelian.

4. Let $\langle G_1, * \rangle \approx \langle G_2, \odot \rangle$. Prove that $\langle G_1, * \rangle$ is abelian if and only if $\langle G_2, \odot \rangle$ is abelian. (*Hint:* Use Exercises 2 and 3.)

†5. Define $\alpha: F_+ \to F$ by $x\alpha = \ln x$ for all $x \in F_+$. Show that α is an isomorphism from $\langle F_+, \cdot \rangle$ onto $\langle F, + \rangle$. ($\ln x$ = natural logarithm of x.)

6. Find two isomorphisms between $\langle Z, + \rangle$ and $\langle E, + \rangle$, where E is the set of even integers.

7. Why is it impossible for $\langle Z_m, +_m \rangle$ and $\langle Z, + \rangle$ to be isomorphic?

†8. There are two isomorphisms between $\langle Z_4, +_4 \rangle$ and $\{1, -1, i, -i\}$ with usual complex multiplication. The last example in this section gives one of them. Find the other one.

Exercises 9 through 12 involve isomorphisms on a group. Let $\text{Aut}(G)$ denote the set of all isomorphisms on $\langle G, * \rangle$. Each element of $\text{Aut}(G)$ is called an *automorphism* of G.

†9. Let g be a fixed element in G. Define $\phi: G \to G$ by $x\phi = g^{-1} * x * g$ for all $x \in G$. Prove that ϕ is an automorphism on $\langle G, * \rangle$.

10. Find all automorphisms on $\langle Z_4, +_4 \rangle$ and verify computationally that, together with composition, they form a group. (*Hint:* There are only two.)

11. Show that $\text{Aut}(G)$, together with composition, forms a group.

12. Let $\eta: G \to G$, defined by $g\eta = g^{-1}$, for all $g \in G$. Prove that $\langle G, * \rangle$ is abelian if and only if $\eta \in \text{Aut}(G)$.

13. Define $\alpha: F \to F$ by $x\alpha = 10x$ for all $x \in F$. Check to see if α is a homomorphism from $\langle F, + \rangle$ into $\langle F, \cdot \rangle$. What is $F\alpha$?

14. Establish an infinite collection of distinct groups, each of which is isomorphic to $\langle Z, + \rangle$. (*Hint:* See Example 2 of this section.)

15. Show that there are only two nonisomorphic groups with four elements by constructing the two possible operation tables.

3.3   PERMUTATION GROUPS

We now present the essentials necessary to establish the equivalency between groups in general and groups of mappings.

3.3.1 DEFINITION. A 1–1 mapping from a nonempty set A onto itself is called a *permutation* on A.

Intuitively, a permutation is an arrangement of the elements of a set in a specified order. Once an initial order is established, a permutation simply specifies another (possibly the same) order. When A

is finite and small, we denote a permutation by listing the "new order" immediately under the "initial order." For example,

$$\alpha = \begin{pmatrix} a & b & c \\ c & b & a \end{pmatrix}$$

is the permutation on $\{a,b,c\}$ that corresponds a to c, b to b, and c to a (i.e., $a\alpha = c$, $b\alpha = b$, and $c\alpha = a$). This listing completely describes the map by placing the image of each element directly beneath the element.

Let $A = \{1,2,3\}$. The permutations on A are

$$E = \begin{pmatrix} 1 & 2 & 3 \\ 1 & 2 & 3 \end{pmatrix}, \quad R = \begin{pmatrix} 1 & 2 & 3 \\ 2 & 3 & 1 \end{pmatrix}, \quad R^2 = \begin{pmatrix} 1 & 2 & 3 \\ 3 & 1 & 2 \end{pmatrix}$$

$$P = \begin{pmatrix} 1 & 2 & 3 \\ 1 & 3 & 2 \end{pmatrix}, \quad S = \begin{pmatrix} 1 & 2 & 3 \\ 3 & 2 & 1 \end{pmatrix}, \quad T = \begin{pmatrix} 1 & 2 & 3 \\ 2 & 1 & 3 \end{pmatrix}$$

This notation enables us to compose these maps rather easily. In order to find the image of an element under composition, just follow the natural path as indicated below:

$$RP = \begin{pmatrix} 1 & 2 & 3 \\ 2 & 3 & 1 \end{pmatrix}\begin{pmatrix} 1 & 2 & 3 \\ 1 & 3 & 2 \end{pmatrix} = \begin{pmatrix} 1 & 2 & 3 \\ 3 & 2 & 1 \end{pmatrix} = S.$$

Using this procedure on all possible combinations of the permutations on A, we obtain the operation table (see 3.3.2) for composition on $\{E,R,R^2,P,S,T\}$.

3.3.2 TABLE

	E	R	R^2	P	S	T
E	E	R	R^2	P	S	T
R	R	R^2	E	S	T	P
R^2	R^2	E	R	T	P	S
P	P	T	S	E	R^2	R
S	S	P	T	R	E	R^2
T	T	S	P	R^2	R	E

Since composition of maps is associative, it is immediately apparent that we have a nonabelian group (if it is not apparent, review 2.6). This group is called the *symmetric group on three symbols* and is denoted $\mathscr{S}_3$.

The permutations in $\mathscr{S}_3$ are frequently called the *rigid motions* of an equilateral triangle. The following discussion explains why. Consider an equilateral triangle as a block in an IQ test which is to be placed into an appropriate hole. There are six different ways this task can be accomplished. Label the vertices in each as indicated.

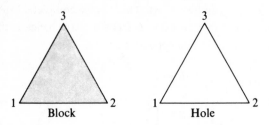

Each permutation on A—that is, each element in $\mathscr{S}_3$—corresponds to one way of fitting the block in the hole. E, the identity map, represents the placement that matches each vertex on the block with one in the hole represented by the same number. R and R^2 correspond to placements that rotate the block 120 and 240 degrees counterclockwise respectively. P, S, and T correspond to placements that flip the block over the dotted lines p, s, and t, respectively, in the diagram.

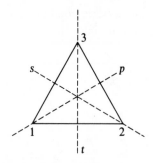

The preceding discussion was based on the collection of permutations on a set with three elements. We generalize now by considering the collection of permutations on an arbitrary set A.

3.3.3 THEOREM. The collection $\mathscr{S}$, of all permutations on a nonempty set A, together with composition is a group.

Proof: Clearly composition is an associative operation on $\mathscr{S}$. The identity map on A, i_A, is the identity in $\mathscr{S}$. For $\alpha \in \mathscr{S}$, α^{-1} exists and belongs to $\mathscr{S}$, since α is both 1–1 and onto. Hence $\mathscr{S}$, together with composition, is a group.▲

The set A in 3.3.3 can be any nonempty set, finite or infinite. When A is finite and has n elements, the associated group of permutations is called the *symmetric group on n symbols* and is denoted $\mathscr{S}_n$.

Permutation groups have been the object of extensive study, the primary reason being an elegant classification theorem due to Arthur Cayley, a nineteenth-century British mathematician.

3.3.4 CAYLEY'S THEOREM. Every group is isomorphic to a group of permutations.

Proof: Let $\langle G,* \rangle$ be an arbitrary group. We will construct a group of permutations on the set G and then show the isomorphism between this new group and $\langle G,* \rangle$.

The construction. Let $g \in G$. Define α_g by $x\alpha_g = x * g$ for all $x \in G$. Clearly $\alpha_g : G \to G$. We now show that α_g is 1–1 and onto. For the 1–1 property, suppose that $x\alpha_g = y\alpha_g$. Then $x * g = y * g$ and $x = y$, which implies that α_g is 1–1. To verify that $G\alpha_g = G$, it suffices to show that $G \subseteq G\alpha_g$, since obviously $G\alpha_g \subseteq G$. Let $y \in G$; then $y = (y * g^{-1}) * g = (y * g^{-1})\alpha_g \in G\alpha_g$. Now $(y * g^{-1}) \in G$, so that α_g is onto. Therefore for each $g \in G$, α_g is a permutation on G. Exercise 3 of this section varifies that $\mathscr{G} = \{\alpha_g \,|\, g \in G\}$ is a group under composition.

The verification that $\mathscr{G}$ is a group is indeed needed, since in general $\mathscr{G}$ is not the group of all permutations on G.

The isomorphism. Define $\beta : G \to \mathscr{G}$ by $g\beta = \alpha_g$ for each $g \in G$. Obviously β is an onto map. To establish the 1–1 property, let $x\beta = y\beta$; that is, let $\alpha_x = \alpha_y$. Then for e, the identity in G, we have $e\alpha_x = e\alpha_y$. Thus $e * x = e * y$ and $x = y$. Finally, β possesses the homomorphism property, since $x,y \in G$ imply $(x * y)\beta = \alpha_{x*y} = \alpha_x\alpha_y = (x\beta)(y\beta)$. Exercise 4 of this section verifies that $\alpha_{x*y} = \alpha_x\alpha_y$.▲

EXERCISES

1. Find eight permutations on $S = \{1,2,3,4\}$ that correspond to the rigid motions of a square and construct an operation table for composition.

2. Find four permutations on $S = \{1,2,3,4\}$ that correspond to the rigid motions of a nonsquare rectangle and construct an operation table for composition.

3. Verify that $\mathscr{G}$, as defined in the proof of 3.3.4, is a group under composition.

†4. Prove that $\alpha_{x \cdot y} = \alpha_x \alpha_y$, where α_g is defined as in the proof of 3.3.4.

5. Construct, as in the proof of Cayley's theorem, a permutation group isomorphic to $\langle Z_4, +_4 \rangle$ and exhibit the isomorphism.

6. Define β_n: $Z \to Z$ by $x\beta_n = x + n$ for each $x \in Z$. Show that $S = \{\beta_n \mid n \in Z\}$ together with composition forms a group.

†7. Prove that $\langle Z, + \rangle$ is isomorphic to the group S of Exercise 6. (*Hint:* Construct an isomorphism similar to the isomorphism used in the proof of Cayley's theorem.)

8. Show that $\mathscr{S}_n$ has $n!$ (n factorial) elements.

3.4   SUBGROUPS

An interesting relationship exists between some of the familiar groups we have considered, such as $\langle Q, + \rangle$ and $\langle F, + \rangle$. $Q \subseteq F$ and the operation $+$ on F, if restricted to Q (i.e., if allowed to act only on elements of Q), is the operation $+$ on Q. This same relationship exists between $\langle Q_0, \cdot \rangle$ and $\langle F_0, \cdot \rangle$ and also between the set of even integers under addition and $\langle Z, + \rangle$.

In general, to establish that the operation $*$ in $\langle G, * \rangle$ is an operation on a nonempty subset H of G, we need only show that $(a * b) \in H$ whenever $a,b \in H$. This implies that $*$ (which is a mapping from $G \times G$ to G), when restricted to H, is a mapping from $H \times H$ to H and thus is an operation on H.

Before continuing this section, we introduce, for convenience, some new notation. Consider $\langle G, * \rangle$. Hereafter we denote $a * b$ by ab (read a times b or the product of a and b) and $\langle G, * \rangle$ by the single letter G. Although we are agreeing to use the notation of ordinary multiplication for the operation in G, it should be remembered that normally the operation is not necessarily ordinary multiplication.

3.4.1 DEFINITION. Let H be a nonempty subset of a group G. H is a *subgroup* of G if H is a group with respect to the operation defined on G restricted to H.

When attempting to verify that H is a subgroup of G, we need to show that (1) the operation on G, when restricted to H, is an operation on H, (2) H has an identity element, and (3) each element in H has an inverse in H. It is not necessary to show that the operation is associative on H. Why?

EXAMPLES

1. $U = \{1, -1, i, -i\}$, the four fourth roots of unity, together with complex multiplication, forms a group. $\{1, -1\}$ with complex multiplication is a subgroup of U. $A = \{i, -i\}$ is not a subgroup of U, since complex multiplication is not an operation on A. Furthermore, A has no identity element.

2. Let G be a group with identity e. G itself is a subgroup of G, and $\{e\}$, together with the operation as in G, is a subgroup of G. Both are referred to as *nonproper subgroups* of G and $\{e\}$ is called the *trivial subgroup* of G.

3. $\langle Q, + \rangle$ is a subgroup of $\langle F, + \rangle$; however, $\langle Q_0, \cdot \rangle$ is not a subgroup of $\langle F, + \rangle$. Why not?

4. The four permutations on $\{1,2,3,4\}$ that correspond to the rigid motions of a nonsquare rectangle form a subgroup of the eight permutations on $\{1,2,3,4\}$ that correspond to the rigid motions of a square (see Exercises 1 and 2 of 3.3).

Notice that in each of the examples the identity for the group is also the identity for the subgroup. The next theorem indicates that such is always the case. We will see later (in the study of rings), however, that there are algebraic systems in which the identity for the large structure is not necessarily the identity for a substructure.

3.4.2 THEOREM. Let H be a subgroup of a group G. Then
1. the identity in H is the identity in G, and
2. if $h \in H$, the inverse of h in H is the inverse of h in G.

Proof: Exercise 1 of this section.▲

Fortunately, we do not have to show that a nonempty subset of a group G satisfies all the group axioms in order to establish that it is a subgroup of G. We can use any one of the following criteria.

3.4.3 THEOREM. Let H be a nonempty subset of the group G. The following are equivalent:

1. H is a subgroup of G,
2. for all $a,b \in H$, we have $ab \in H$ and $a^{-1} \in H$,
3. for all $a,b \in H$, we have $ab^{-1} \in H$.

Proof: Clearly (1) implies (2) and (2) implies (3). We will show that (3) implies (1).

The subset H is nonempty; thus there is an element $h \in H$. By (3), $hh^{-1} = e \in H$. For arbitrary $a \in H$, again by (3), $ea^{-1} = a^{-1} \in H$. The operation of G is an operation on H, since for $a,b \in H$ we have $b^{-1} \in H$ and thus $a(b^{-1})^{-1} = ab \in H$. Finally, associativity follows from associativity on G, and we conclude that H is a subgroup of G.▲

The next theorem shows that an even simpler criterion can be used to verify that H is a subgroup if we know that H is a finite set.

3.4.4 THEOREM. If H is a finite nonempty subset of the group G such that for each $a,b \in H$ we have $ab \in H$, then H is a subgroup of G.

Proof: Since H is finite, we can write $H = \{h_1, h_2, \ldots, h_n\}$. By 3.4.3.2, we need to show that $h_k^{-1} \in H$ for arbitrary $h_k \in H$.

Consider $H' = \{h_k h_1, h_k h_2, \ldots, h_k h_n\}$. Via the left cancellation law in G, $h_k h_i = h_k h_j$ if and only if $i = j$; therefore H' has n elements. Now $h_k h_i \in H$ for $1 \leq i \leq n$ by hypothesis. Hence $H' \subseteq H$; and since they have the same number of elements, $H' = H$. There exists $h_r \in H$ such that $h_k h_r = h_k$ since $H' = H$. Now $h_k h_r = h_k e$ in G; and by left cancellation, $h_r = e$. Similarly, there exists $h_t \in H$ so that $h_k h_t = h_r = e$. Therefore h_t is a right inverse of h_k; and, by 3.1.2.2, h_t is the inverse of h_k. Thus $h_k^{-1} = h_t \in H$.▲

For an arbitrary group G, there are certain subsets that are always subgroups. Some examples of such subsets appear below.

3.4.5 DEFINITION. Let α be a homomorphism from a group G to a group G' and let e' be the identity in G'. The *kernel* of α is the set of all pre-images of e' in G. Symbolically,

$$\ker \alpha = \{g \mid g \in G \text{ and } g\alpha = e'\}.$$

3.4.6 THEOREM. If α is a group homomorphism, then ker α is a subgroup of dom α.

Proof: Exercise 2 of this section. ▲

3.4.7 DEFINITION. The *center* of a group G is the set of all elements in G that commute with every g in G. Symbolically,
$$C(G) = \{x \mid x \in G \text{ and } xg = gx \text{ for all } g \in G\}.$$

3.4.8 THEOREM. Let G be a group. $C(G)$ is a subgroup of G.

Proof: Exercise 3 of this section. ▲

EXERCISES

1. Prove 3.4.2.
2. Prove 3.4.6.
3. Prove 3.4.8.
4. Let $\mathscr{H}$ be a nonempty collection of subgroups of a group G. Show that $\cap \mathscr{H}$ is a subgroup of G.
5. Let H be a subgroup of a group G. For $g \in G$, prove that $g^{-1}Hg$ is a subgroup of G, where $g^{-1}Hg = \{g^{-1}hg \mid h \in H\}$.
6. Verify Example 4 of this section.
†7. Let α be a homomorphism from a group G to a group G' and let H' be a subgroup of G'. Prove that H is a subgroup of G and that ker $\alpha \subseteq H$, where $H = \{g \mid g \in G \text{ and } g\alpha \in H'\}$.
†8. Let α be a homomorphism from a group G onto a group G'. Show that α is an isomorphism if and only if ker $\alpha = \{e\}$, where e is the identity in G.
9. In Example 4 of 3.2, find ker η.
10. In Exercise 6 of 3.2, find the kernel for each of the isomorphisms.
11. Show that the group G is abelian if and only if $C(G) = G$.
12. Show that $\mathscr{R} = \{E,R,R^2\}$ is a subgroup of $\mathscr{S}_3$ (see 3.3.2). Find $C(\mathscr{S}_3)$ and $C(\mathscr{R})$.

3.5 COSETS

This section introduces a combination of the second and third techniques mentioned in Section 3.2 by considering a collection of appropriate subsets that partition an arbitrary group.

3.5.1 DEFINITION. Let G be a group, H a subgroup of G, and $g \in G$. Then

1. $gH = \{gh \mid h \in H\}$ is called the *left coset* of g and H in G and
2. $Hg = \{hg \mid h \in H\}$ is called the *right coset* of g and H in G.

The element g is called a *left coset representative* in (1) and a *right coset representative* in (2).

Our goal is to show that the collection of all left cosets of a given subgroup H of a group G is a partition of G and thus, by 2.2.6, determines an equivalence relation on G.

3.5.2 TABLE

	e	a	b	c	d	f	g	h
e	e	a	b	c	d	f	g	h
a	a	b	c	e	g	h	f	d
b	b	c	e	a	f	d	h	g
c	c	e	a	b	h	g	d	f
d	d	h	f	g	e	b	c	a
f	f	g	d	h	b	e	a	c
g	g	d	h	f	a	c	e	b
h	h	f	g	d	c	a	b	e

Let G be the group whose operation is defined in 3.5.2. Let $H = \{e,d\}$. H is a subgroup of G, and the left and right cosets of H are

$$eH = H \qquad He = H$$
$$aH = \{a,g\} \qquad Ha = \{a,h\}$$
$$bH = \{b,f\} \qquad Hb = \{b,f\}$$
$$cH = \{c,h\} \qquad Hc = \{c,g\}$$
$$dH = H \qquad Hd = H$$
$$fH = bH \qquad Hf = Hb$$
$$gH = aH \qquad Hg = Hc$$
$$hH = cH \qquad Hh = Ha.$$

In each case, the coset representative is an element of the coset that it determines; and if two elements belong to the same coset, each determines that particular coset. Moreover, each coset has the same number of elements, and the collection of all left cosets of H partitions G, as does the collection of all right cosets of H.

Since $cH \neq Hc$, a given element does not necessarily determine the same left and right coset. However, for the subgroup $K = \{e,a,b,c\}$ it is easily verified that each element in G does determine the same left and right coset—that is, $eK = Ke$, $bK = Kb$, $aK = Ka$, etc. Subgroups having this property will be studied in Section 4.1.

3.5.3 THEOREM. Let G be a group, H a subgroup of G, and $a,b \in G$. The following statements are equivalent.

1. a and b belong to the same left coset of H in G (i.e., $a \in bH$).
2. a and b determine the same left coset of H in G (i.e., $aH = bH$).
3. $a^{-1}b \in H$.

Proof: (1) implies (2). Let $a \in bH$. Then there exists $h_1 \in H$ such that $a = bh_1$ (and $ah_1^{-1} = b$). If $x \in aH$, then $x = ah_2$ for a suitable $h_2 \in H$. Now $aH \subseteq bH$ since $x = ah_2 = (bh_1)h_2 = b(h_1h_2) \in bH$ (Why?). Let $y \in bH$; then $y = bh_3$ for some $h_3 \in H$. $y = bh_3 = (ah_1^{-1})h_3 = a(h_1^{-1}h_3) \in aH$, and $bH \subseteq aH$. Hence $aH = bH$.

(2) implies (3). Let $aH = bH$. For $x \in aH = bH$ there exist $h_1,h_2 \in H$ such that $x = ah_1$ and $x = bh_2$. Now $bh_2 = ah_1$ and $a^{-1}b = h_1h_2^{-1} \in H$.

(3) implies (1). Let $a^{-1}b \in H$. Then there exists $h_1 \in H$ such that $a^{-1}b = h_1$. Let $a \in xH$; then for some $h_2 \in H$, $a = xh_2$ and $b = ah_1 = (xh_2)h_1 = x(h_2h_1) \in xH$. Hence a and b are elements of the same coset of H in G.▲

3.5.4 COROLLARY. Let G be a group and H a subgroup of G. For $a,b \in G$, either $aH = bH$ or $aH \cap bH = \emptyset$.

Proof: Exercise 2 of this section.▲

3.5.5 THEOREM. Let G be a group and H a subgroup of G. The collection $\mathscr{P}$ of left cosets of H in G is a partition of G.

Proof: $\mathscr{P} = \{gH \,|\, g \in G\}$.
1. Let $gH \in \mathscr{P}$. Then $g = ge \in gH$, so $gH \neq \emptyset$.
2. Since $g \in gH$, each $g \in G$ belongs to some left coset (i.e., $G \subseteq \cup \mathscr{P}$), and by definition each $aH \subseteq G$. Thus $G = \cup \mathscr{P}$.
3. By 3.5.4, the cells in $\mathscr{P}$ are pairwise disjoint.▲

EXERCISES

1. Restate and prove 3.5.3 for right cosets of H in G.

2. Prove 3.5.4.

3. Show that the group $G = \{e,a,b,c,d,f,g,h\}$ (the operation is defined in 3.5.2) is isomorphic to the group constructed in Exercise 1 of Section 3.3.

4. List the right and left cosets of $K = \{e,a,b,c\}$ (the operation is defined in 3.5.2).

†5. Let H be a subgroup of a group G. Show that the only left coset of H in G, which is also a subgroup of G, is H itself.

†6. Let H be a subgroup of a group G and $g \in G$. Prove that $gH = H$ if and only if $g \in H$.

7. Let H be a subgroup of a group G. Define a relation $\sim$ on G as follows: For all $a, b \in G$, $a \sim b$ if and only if $a^{-1}b \in H$. Show that $\sim$ is an equivalence relation on G.

8. (From Exercise 7) Prove that the equivalence classes determined by $\sim$ are the left cosets of H in G, thus giving an alternate proof of 3.5.5.

3.6   THE THEOREM OF LAGRANGE

The partition $\mathscr{P}$ determined in 3.5.5 is extremely fruitful in group theory. Often valuable information about the group G can be obtained by considering $\mathscr{P}$ in some particular way. In Section 4.2 we will investigate $\mathscr{P}$ in connection with group homomorphisms. For the time being, this partition will be considered as a decomposition of a group G that has only a finite number of elements.

3.6.1 DEFINITION. The *order* of a group G, denoted $|G|$, is the number of elements in the set G. If $|G| = n$ for some positive integer n, then G is called a *finite group* or a group of order n. Otherwise G is called an *infinite group*.

$\mathscr{S}_3$ is a group of order 6, while $\langle Z,+\rangle$ and $\langle F_+,\cdot\rangle$ are infinite groups. The congruence relations on Z enable us to construct finite groups of every order.

3.6.2 THEOREM. There exists at least one group of order n for each positive integer n.

Proof: $\langle Z_n, +_n \rangle$ is a group and $|Z_n| = n$ for each positive integer n. ▲

3.6.3 DEFINITION. Let H be a subgroup of a group G. The *index* of H in G, denoted $[G:H]$, is the number of distinct left cosets of H in G.

The symbol $[G:H]$ also denotes the number of distinct right cosets of H in G. This point is easily verified by constructing a 1-1 correspondence between the set of all left cosets and the set of all right cosets (see Exercise 8 of this section). Notice that we do not require G to be finite in 3.6.3, nor do we require $[G:H]$ to be finite. In the following examples, Example 2 exhibits a subgroup of finite index in an infinite group and Example 3 exhibits a subgroup of infinite index in an infinite group. Certainly when $|G|$ is finite, $[G:H]$ is finite.

EXAMPLES

1. G with the operation defined in 3.5.2 is a group. $H = \{e,d\}$ and $K = \{e,a,b,c\}$ are subgroups of G. $[G:H] = 4$ and $[G:K] = 2$.

2. Consider $\langle Z,+ \rangle$ and $\langle E,+ \rangle$, where E is the set of even integers. The left cosets of E in Z are $0 + E = E$ and $1 + E$ (here the symbol $+$ is used for the operation rather than juxtaposition). Thus $[Z:E] = 2$.

3. Consider $\langle Z,+ \rangle$ as a subgroup of $\langle Q,+ \rangle$. For each $n \in Z_+$, the left cosets $1/n + Z$ are distinct. This is by no means a complete list of the left cosets of Z in Q; however, it is enough to conclude that $[Q:Z]$ is infinite.

It is interesting to note that each left coset of the subgroup H in G contains the same number of elements as H (in the sense of 1-1 correspondence when H is infinite). When G is finite, this leads directly to a classical result due to the French mathematician Joseph L. Lagrange (1736–1831).

3.6.4 LEMMA. Let H be a subgroup of a group G. Each left coset of H in G contains $|H|$ elements.

Proof: Let gH be a left coset of H in G. The mapping $\alpha: H \to gH$ defined by $h\alpha = gh$ for each $h \in H$ is a 1–1 correspondence. ▲

3.6.5 LAGRANGE'S THEOREM. If H is a subgroup of a finite group G, then $|G| = |H| \cdot [G:H]$.

Proof: The partition $\mathscr{P} = \{gH \mid g \in G\}$ of G contains $[G:H]$ distinct cells. Each cell contains $|H|$ elements by 3.6.4. The cells of $\mathscr{P}$ are pairwise disjoint and their union is G. Therefore $|G| = |H| \cdot [G:H]$. ▲

This theorem tells us that the order of every subgroup of a finite group must be a divisor of the order of the group. For instance, a group of order 6 cannot have a subgroup of order 4, because 4 does not divide 6 (with a zero remainder). A subgroup of a group of order 6 must be of order 1, 2, 3, or 6. On the other hand, 3.6.5 does not assure us that subgroups of a particular order exist just because the number is a divisor of the order of the group.

EXERCISES

†1. Let S be the set of integral multiples of 6; that is, $S = \{\ldots, -12, -6, 0, 6, 12, \ldots\}$. $\langle S, + \rangle$ is a subgroup of both $\langle Z, + \rangle$ and $\langle E, + \rangle$, where E is the set of even integers. Calculate $[Z:S]$ and $[E:S]$.

†2. HK is defined to be $\{hk \mid h \in H$ and $k \in K\}$, where H and K are subsets of a group G. If H and K are different subgroups of $\mathscr{S}_3$, each of order 2, show, using Lagrange's theorem, that HK is not a subgroup of $\mathscr{S}_3$ (see 3.3.2).

3. Let H and K be subgroups of a group G and let HK be defined as in Exercise 2. Prove that HK is a subgroup of G if and only if $HK = KH$.

4. Prove that α in 3.6.4 is a 1–1 correspondence.

5. Determine $[\mathscr{S}_3 : C(\mathscr{S}_3)]$. See Exercise 12 in Section 3.4.

6. Let G be a finite group. If H is a subgroup of G and K is a subgroup of H, show that $[G:K] = [G:H][H:K]$.

7. Restate and prove 3.6.4 for right cosets of H in G.

†8. Let H be a subgroup of a group G. Show that $[G:H]$ also denotes the number of distinct right cosets of H in G.

3.7   CYCLIC GROUPS

In this section we deal primarily with "powers" or "multiples" of elements in a group. We adopt the usual notation for these concepts.

3.7.1 DEFINITION. Let g be an element of a group G with identity e.

    1. $g^0 = e$.

    2. $g^m = gg \ldots g$ (m factors) for m a positive integer.

    3. $g^{-m} = (g^{-1})^m$ for m a positive integer.

In an additive group (where the operation is denoted by $+$ rather than juxtaposition), the definition for powers becomes a definition for multiples.

3.7.1' DEFINITION. Let g be an element of an additive group G with identity e.

    1. $0g = e$.

    2. $mg = g + g + \cdots + g$ (m addends) for m a positive integer.

    3. $(-m)g = m(-g)$ for m a positive integer (recall that the inverse of g is denoted $-g$ in an additive group).

A necessary distinction to be made is that m and 0 are integers, not elements of the group G (unless $Z \subseteq G$ or $G \subseteq Z$). In part (2), we are not "multiplying" m and g, we are simply using mg to denote the sum of m g's.

With the aid of the following lemma, we can justify the familiar laws of exponents.

3.7.2 LEMMA. Let $m \in Z_+$ and g be an element of a group G. Then

    1. $(g^m)^{-1} = (g^{-1})^m$ and

    2. g to the negative one power is g inverse.

Proof: Exercise 1 of this section.▲

Part (2) of 3.7.2 tells us that g^{-1} is not an ambiguous symbol. The two possible interpretations are equivalent and can be used interchangeably.

3.7.3 THEOREM. Let G be a group, $g \in G$ and $m,n \in Z$. Then

    1. $g^m g^n = g^{m+n}$ and

    2. $(g^m)^n = g^{mn}$.

Proof: 1. Case 1. Let $m, n \in Z_+$. Then
$$g^m g^n = \underbrace{(gg \ldots g)}_{m \text{ factors}} \underbrace{(gg \ldots g)}_{n \text{ factors}} = g^{m+n}.$$
$$\underbrace{}_{m + n \text{ factors}}$$

Case 2. For $m = 0$ or $n = 0$, the result follows from $g^0 = e$ and the additive property of the integer 0.

Case 3. Let m, n be negative integers. Then for $p = -m$ and $q = -n$, we have $g^m g^n = g^{-p} g^{-q} = (g^{-1})^p (g^{-1})^q = (g^{-1})^{p+q} = g^{-(p+q)}$ $= g^{-p-q} = g^{m+n}$.

Case 4. Let m be positive and n negative. Then for $p = -n$, we have

$$g^m g^n = g^m g^{-p} = g^m (g^{-1})^p = \underbrace{(gg \ldots g)}_{m \text{ factors}} \underbrace{(g^{-1} g^{-1} \ldots g^{-1})}_{p \text{ factors}} = \begin{cases} g^{m-p} & \text{if } m \geq p \\ (g^{-1})^{p-m} & \text{if } m < p \end{cases}$$

$$= g^{m-p} \text{ (in both cases)}$$
$$= g^{m+n}.$$

Case 5. Let m be negative and n positive. Interchange the roles of m and n in Case 4.

2. Case 1. Let $n = 0$. Clearly $(g^m)^n = e = g^{mn}$.

Case 2. Let $n \in Z_+$. Then
$$(g^m)^n = \underbrace{g^m g^m \ldots g^m}_{n \text{ factors}} = g^{\overbrace{m+m+\cdots+m}^{n \text{ addends}}} = g^{mn}.$$

Case 3. Let n be a negative integer. Then for $p = -n$
$$(g^m)^n = (g^m)^{-p} = \underbrace{[(g^m)^{-1}]^p = [(g^{-1})^m]^p}_{\text{by 3.7.2}} = (g^{-1})^{mp} = g^{-mp} = g^{mn}. \blacktriangle$$

These properties lead to an important and interesting class of groups via the following subgroups.

3.7.4 DEFINITION. Let g be an element of a group G. The subset of G consisting of all integral powers of g is called the *cyclic subgroup generated by g.* Symbolically,
$$\langle g \rangle = \{g^n \mid n \in Z\}.$$

It is easily verified that $\langle g \rangle$ is indeed a subgroup of G. The most important property of $\langle g \rangle$ is that each of its elements can be expressed as an appropriate power of g (or appropriate multiples if the operation uses additive notation).

Consider the following groups:

$Z_4 = \{\bar{0},\bar{1},\bar{2},\bar{3}\}$

$+_4$	$\bar{0}$	$\bar{1}$	$\bar{2}$	$\bar{3}$
$\bar{0}$	$\bar{0}$	$\bar{1}$	$\bar{2}$	$\bar{3}$
$\bar{1}$	$\bar{1}$	$\bar{2}$	$\bar{3}$	$\bar{0}$
$\bar{2}$	$\bar{2}$	$\bar{3}$	$\bar{0}$	$\bar{1}$
$\bar{3}$	$\bar{3}$	$\bar{0}$	$\bar{1}$	$\bar{2}$

$K = \{e,a,b,c\}$

	e	a	b	c
e	e	a	b	c
a	a	e	c	b
b	b	c	e	a
c	c	b	a	e

The cyclic subgroups of Z_4 and K are

$$\langle\bar{0}\rangle = \{\bar{0}\} \qquad \langle e\rangle = \{e\}$$
$$\langle\bar{1}\rangle = \{\bar{0},\bar{1},\bar{2},\bar{3}\} \qquad \langle a\rangle = \{e,a\}$$
$$\langle\bar{2}\rangle = \{\bar{0},\bar{2}\} \qquad \langle b\rangle = \{e,b\}$$
$$\langle\bar{3}\rangle = \{\bar{0},\bar{1},\bar{2},\bar{3}\} \qquad \langle c\rangle = \{e,c\}.$$

In Z_4, $\langle\bar{1}\rangle = \langle\bar{3}\rangle$, so the generator of a cyclic subgroup is not necessarily unique. One notable difference between Z_4 and K is that Z_4 can be considered as a cyclic subgroup of itself, since $Z_4 = \langle\bar{1}\rangle$, whereas K does not have this property.

3.7.5 DEFINITION. A group G that can be expressed as a cyclic subgroup of itself is called a *cyclic group*. Equivalently, G is cyclic if there exists $g \in G$ such that $\langle g\rangle = G$.

EXAMPLES

1. $\langle Z_n,+_n\rangle$ for each positive integer n is cyclic.
2. $\{1,-1,i,-i\}$ with complex multiplication is a cyclic group with generators i and $-i$.
3. $\langle Z,+\rangle$ is a cyclic group with generators 1 and -1.
4. $\{\bar{1}\}$ is the cyclic subgroup generated by $\bar{1}$ when the overriding group is $\langle Z_5\backslash\{0\},\cdot_5\rangle$. However, $\langle\bar{1}\rangle = Z_5$ when considered as a subgroup of $\langle Z_5,+_5\rangle$.

There is a difference between a cyclic group and a cyclic subgroup of a group. Every group has cyclic subgroups, but not every group is cyclic. Some groups, like K in the discussion following 3.7.4, do not possess "big enough" cyclic subgroups to qualify as cyclic groups.

Since noncyclic groups exist, we suspect that an arbitrary group might contain noncyclic subgroups. Most often this is the case; however, when G is cyclic, we arrive at the conclusion in the following theorem.

3.7.6 THEOREM. Every subgroup of a cyclic group is cyclic.

Proof: Let $G = \langle g \rangle$ and let H be a subgroup of G. If $H = \{e\}$, where e is the identity in G, then $H = \langle e \rangle$. Suppose that $H \neq \{e\}$. Then there exists $x \in H$ such that $x \neq e$. $G = \langle g \rangle$ and $x \in G$, so $x = g^s$ for some integer s. Without loss of generality, we may assume that $s > 0$ (for $s < 0$ we would use x^{-1}, which is also in H). The collection of all positive exponents $p \leq s$ so that $g^p \in H$ is a nonempty subset of Z_+. This set has a smallest element, say m, since Z_+ is well ordered. Therefore m is the smallest positive integer such that $g^m \in H$. We now show that $H = \langle g^m \rangle$.

Let $h \in H$. Then $h = g^n$ for some $n \in Z$. By the division algorithm (2.7.1), there exist integers q and r such that $n = qm + r$, where $0 \leq r < m$. Now $g^m \in H$; thus $(g^{-m})^q \in H$ and $g^n(g^{-m})^q \in H$. Furthermore, $g^r = g^{qm+r} g^{-qm} = g^n g^{-qm} = g^n(g^{-m})^q \in H$. Since m is the smallest positive integer such that $g^m \in H$ and r is a non-negative integer with $g^r \in H$, we must have $r = 0$. Therefore $h = g^n = g^{qm} = (g^m)^q \in \langle g^m \rangle$ and $H \subseteq \langle g^m \rangle$. Now since $g^m \in H$, we also have $\langle g^m \rangle \subseteq H$ and equality holds.▲

Recall that in the discussion of Lagrange's theorem we indicated that the order of a subgroup must divide the order of a group; yet the fact that a number m divides the order of a group does not mean that a subgroup of order m exists. If we require the group to be cyclic, however, not only does a subgroup of order m exist but it is also unique. The remainder of this section proceeds to establish this result.

3.7.7 DEFINITION. Let G be a group and $g \in G$. The order of the group generated by g, $|\langle g \rangle|$, is called the *order* of g and is denoted $|g|$.

If $|G|$ is finite, then $|g|$ for each $g \in G$ is clearly finite. However, when $|G|$ is infinite, $|g|$ may be either finite or infinite.

EXAMPLES

1. In $\langle Z, + \rangle$, $|0| = 1$ and $|1|$ is infinite.

2. In $\langle Q_0, \cdot \rangle$, we have $|1| = 1$, $|-1| = 2$, and every other element has infinite order.

3. For $\langle P(S), + \rangle$ from Exercise 11 in Section 3.1, each element different from the identity $\emptyset$ has order 2 since $A + A = (A \cup A) \setminus (A \cap A) = \emptyset$. When S is an infinite set, this becomes an example of an infinite group in which every element has finite order.

3.7.8 THEOREM. Let g be an element of finite order in a group G. Then $|g| = n$ if and only if n is the smallest positive integer such that $g^n = e$, where e is the identity in G.

Proof: Let n be the smallest positive integer such that $g^n = e$. Suppose that $g^i = g^j$ for some integers i, j with $0 \leq j < i < n$. Then $g^{i-j} = e$ and $i - j < n$. This result contradicts the minimality of n; therefore $g^i \neq g^j$ for all non-negative integers $i, j < n$. That is, $g^0, g^1, \ldots, g^{n-1}$ are distinct. Hence $|g| \geq n$. Let t be an integer different from $0, 1, 2, \ldots, n - 1$. Then there exist integers q and r such that $t = nq + r$ with $0 \leq r < n$. Now $g^t = g^{nq+r} = (g^n)^q g^r = (e)^q g^r = g^r$, and we conclude that $\langle g \rangle = \{g^0, g^1, \ldots, g^{n-1}\}$; that is, $|g| = n$.

Conversely, assume that $|g| = n$. Then not all positive powers of g are distinct; that is, there exist integers i, j such that $0 < i < j$ and $g^i = g^j$. Now $g^{j-i} = e$ and $j - i > 0$; thus there is a positive power of g equal to e. Let m be the smallest positive integer such that $g^m = e$. By the first part of the theorem, $|g| = m$; therefore $m = n$. ▲

3.7.9 COROLLARY. Let G be a group. If $g \in G$ and $|g| = n$, then $\langle g \rangle = \{g^0, g^1, \ldots, g^{n-1}\}$.

Proof: Contained in the proof of 3.7.8. ▲

3.7.10 COROLLARY. Let G be a group with identity e. If $|G| = n$, then $g^n = e$ for all $g \in G$.

Proof: Exercise 4 of this section. ▲

We conclude this section with a complete characterization of the subgroup structure of a finite cyclic group. This result includes a partial converse of Lagrange's theorem.

3.7.11 THEOREM. A cyclic group G of order n has a unique subgroup of order m if and only if m divides n and $m \in Z_+$.

Proof: If G has a subgroup of order m, then m divides n by Lagrange's theorem.

Conversely, suppose that m divides n and $m > 0$. Then there is a positive integer s such that $sm = n$. Let $G = \langle g \rangle$. Consider $\langle g^s \rangle$. $(g^s)^m = g^{sm} = g^n = e$ by 3.7.10. Suppose, by way of contradiction, that there is a positive integer $r < m$ such that $(g^s)^r = e$. Then $sr < sm = n$ and $g^{sr} = e$, which contradicts the minimality of n as guaranteed by 3.7.8. Hence m is the smallest positive integer such that $(g^s)^m = e$; that is, $|g^s| = m$ and $H = \langle g^s \rangle$ is a subgroup of order m.

In order to show that there is exactly one subgroup of order m, let H and K be subgroups of order m with $H = \langle g^s \rangle$ as above. K is cyclic (Why?), so that $K = \langle g^t \rangle$. There exist integers q and r such that $t = sq + r$ with $0 \le r < s$. $g^t = g^{sq+r}$ and $(g^t)^m = g^{sqm}g^{rm}$, but $g^{tm} = e$ and $(g^{sm})^q = e$. Hence $g^{rm} = e$. Now $rm < sm = n$ (since $r < s$) and $g^{rm} = e$ imply $rm = 0$, since n is the smallest positive integer such that $g^n = e$. Furthermore, $rm = 0$ implies $r = 0$ and $t = sq + r = sq$. Finally, $g^t = (g^s)^q \in H$ and $K = H$. ▲

To show that the group in 3.7.11 must be cyclic in order for the conclusion to follow, consider $\langle P(S), + \rangle$, where $S = \{1,2,3\}$. We noted in Example 7 of this section that each element in $P(S)$, other than the identity $\emptyset$, is of order 2. Therefore $P(S)$ is not cyclic, $|P(S)| = 8$, and there are seven subgroups of order 2. Furthermore, if $|G| = n$ and m divides n, it is possible that G does not have a subgroup of order m. For example, $\mathscr{S}_4$ has a subgroup of order 12, which does not have a subgroup of order 4.

EXERCISES

1. Prove 3.7.2.
2. Verify Cases 2 and 5 in part (1) of the proof of 3.7.3.
3. Show that $\langle g \rangle$ is a subgroup of a group G for $g \in G$.

4. Prove 3.7.10.

†5. Prove that every cyclic group is abelian.

6. Find all cyclic subgroups of $\mathscr{S}_3$.

7. Prove $\langle Z_n, +_n \rangle$ is cyclic for each positive integer n.

†8. Let G be a cyclic group with generator g. Let ϕ be a homomorphism from G onto a group G'. Show that $G' = \langle g\phi \rangle$.

A cyclic group consists of powers of a single element. Since the law of exponents holds ($g^n g^m = g^{n+m}$), we might expect the operation of a cyclic group to be quite analogous to the operation of addition on Z. Such is indeed the case. The extent of this analogy is shown in Exercises 9 and 10.

9. If G is a cyclic group of order n, prove that $G \approx \langle Z_n, +_n \rangle$.

10. If G is an infinite cyclic group, show that $G \approx \langle Z, + \rangle$.

11. Use Exercises 9 and 10 to show that cyclic groups of the same order are isomorphic.

†12. An integer $p > 1$ is *prime* if the only positive integral divisors of p are 1 and p. Show, using Lagrange's theorem, that every group of prime order is cyclic.

13. Let A and B be subgroups of a group G with $|A|$ prime (see Exercise 12) and $\{e\}$ properly contained in $A \cap B$. Show that $A \subseteq B$.

†14. Let G be a group. Show that G is abelian if and only if $a^2 b^2 = (ab)^2$ for all $a, b \in G$.

15. Let G be a group. Prove that G is abelian if $|g| = 2$ for all $g \in G$ different from the identity.

16. Let $\alpha: G \rightarrow G'$ be an onto group homomorphism with G a finite group. For each $g' \in G'$, prove that $|g'|$ divides $|g|$, where $g \in G$ and $g\alpha = g'$.

17. (From Exercise 16) If α is an isomorphism, show that $|g'| = |g|$.

18. Let a and ab be elements of finite order in a group G. Show that $|a| = |a^{-1}|$, $|a| = |g^{-1} ag|$ for all $g \in G$, and $|ab| = |ba|$.

19. Let G be a cyclic group with generator g. Show that g^{-1} is also a generator for G.

20. Show that the order of every nontrivial subgroup of Z is infinite.

†21. Show that every nonabelian group contains proper subgroups.

22. Let G be a nonabelian group of order pq, where p and q are distinct primes (see Exercise 12). Show that G contains a subgroup of order p or a subgroup of order q.

23. Give an example of a noncyclic group in which every proper subgroup is cyclic.

24. Let G be a group of order pq, where p and q are primes. If G contains, at most, p elements of order p and, at most, q elements of order q, show that G is cyclic. (*Hint:* Use the fact that the order of an element divides the order of the group.)

CHAPTER 4

MORE ON GROUPS

This chapter examines the close relationship between the study of groups via subgroups and the study of groups via homomorphisms. In fact, it exposes an interrelationship between all three of the basic techniques that we have been using to investigate groups (see 3.2). It concludes with a reference section that presents many of the group theoretical concepts in additive notation. This section is included to help simplify the transition to the study of algebraic systems with more than one operation.

4.1   NORMAL SUBGROUPS

Section 3.5 established that some subgroups of a group G have the property that each element in G determines the same left and right coset. In this section, we associate each homomorphism of a group with a subgroup possessing this property.

4.1.1 DEFINITION. Let H be a subgroup of the group G. H is called a *normal subgroup* of G if for each $g \in G$ we have $gH = Hg$.

We immediately note that $\{e\}$ and G are normal subgroups of G. It may happen that these are the only normal subgroups of G, in which case we say that G is a *simple group*.

EXAMPLES

1. For the group G given in 3.5.2, H is not a normal subgroup of G, whereas K is. Is G simple? $M = \{e,b\}$ is a subgroup of G. Is M normal in G? M is also a subgroup of K; is M normal in K?

2. Every subgroup of $\langle Z_n, +_n \rangle$ is normal. When n is a prime, Z_n is a simple group. Why isn't Z_n simple when n is not a prime?

The following theorem establishes a choice of methods for verifying the normality of a given subgroup. Notationally, $g^{-1}Hg$ is used to represent $\{g^{-1}hg \mid h \in H\}$, where g is a fixed element in G.

4.1.2 THEOREM. Let H be a subgroup of the group G. The following are equivalent.

1. H is normal in G (i.e., $gH = Hg$ for each $g \in G$).
2. $gH \subseteq Hg$ for each $g \in G$.
3. $g^{-1}Hg \subseteq H$ for each $g \in G$.
4. $g^{-1}Hg = H$ for each $g \in G$.
5. $g^{-1}hg \in H$ for each $g \in G$ and each $h \in H$.

Proof: (1) implies (2). Obvious.

(2) implies (3). Assume that $gH \subseteq Hg$ for all $g \in G$. Let $x \in G$ and $y \in x^{-1}Hx$. Then for an appropriate $h_1 \in H$, we have $y = x^{-1}h_1 x$. Since $x^{-1}H \subseteq Hx^{-1}$, there is an $h_2 \in H$ such that $x^{-1}h_1 = h_2 x^{-1}$. Therefore $y = x^{-1}h_1 x = h_2 x^{-1}x = h_2 \in H$ and $x^{-1}Hx \subseteq H$.

(3) implies (4). Let $x \in G$. We have $x^{-1}Hx \subseteq H$ and $xHx^{-1} \subseteq H$. Let $h \in H$. Then $h = x^{-1}(xhx^{-1})x = x^{-1}h'x \in x^{-1}Hx$. This implies $H \subseteq x^{-1}Hx$. Hence $x^{-1}Hx = H$.

(4) implies (5). Obvious.

(5) implies (1). Let $x \in G$. For $y \in xH$, we have $y = xh_1$ for a suitable $h_1 \in H$. By (5), there is an $h_2 \in H$ such that $xh_1x^{-1} = h_2$; thus $yx^{-1} = (xh_1)x^{-1} = h_2$. Hence $y = h_2x$ and $xH \subseteq Hx$. Similarly, $Hx \subseteq xH$ and H is normal in G.▲

Note that (1) is equivalent to (2) and that (3) is equivalent to (4). In general, in order to show set equality, we must verify containment in two directions. However, for the cosets gH and Hg, verifying containment in one direction is sufficient.

The next theorem accomplishes the goal of this section. It associates with each homomorphism, α of G, a normal subgroup of G; namely, ker α.

4.1.3 THEOREM. If $\alpha: G \to G'$ is a group homomorphism, then ker α is a normal subgroup of G.

Proof: By 3.4.6, ker α is a subgroup of G. To show normality, we appeal to 4.1.2.5. Let $g \in G$ and $h \in$ ker α. Then we have $(g^{-1}hg)\alpha = (g^{-1}\alpha)(h\alpha)(g\alpha) = (g\alpha)^{-1}e'(g\alpha) = (g\alpha)^{-1}(g\alpha) = e'$, where e' is the identity in G'. $(g^{-1}hg)\alpha = e'$ implies $g^{-1}hg \in$ ker α. Therefore ker α is normal in G.▲

We will show in 4.2.8 that subgroups of the form ker α for some group homomorphism α are the only normal subgroups of G.

EXERCISES

†1. Let G be a group. Show that $C(G)$ is a normal subgroup of G. See 3.4.7 for the definition of $C(G)$.

2. Let G be an abelian group. Show that every subgroup of G is normal in G.

3. Prove that a cyclic group G is simple if and only if $|G|$ is prime.

†4. If $\alpha: G \to G'$ is an onto group homomorphism and H is a normal subgroup of G, prove that $H\alpha$ is a normal subgroup of G'.

5. Let $\alpha: G \to G'$ be an onto group homomorphism and H' be a normal subgroup of G'. Prove that $H = \{h \in G \mid h\alpha \in H'\}$ is a normal subgroup of G and ker $\alpha \subseteq H$.

6. Find all normal subgroups of $\mathscr{S}_3$.

7. If H and K are normal subgroups of a group G, show that $H \cap K$ is normal in G.

8. Let H and K be subgroups of a group G. Prove that HK is a subgroup of G if either H or K is normal in G. Recall that $HK = \{hk \mid h \in H$ and $k \in K\}$.

†9. If H and K are normal subgroups of the group G, show that HK is a normal subgroup in G.

10. If H is a normal subgroup of a group G and K is any subgroup of G, prove that $H \cap K$ is a normal subgroup of K.

11. Let H be a subgroup of a group G. If $[G:H] = 2$, prove that H is normal in G. (*Hint:* Consider the collection of left cosets and the collection of right cosets.)

12. Show that the relation "is a normal subgroup of" on a nonempty collection of groups is reflexive but not symmetric or transitive. (*Hint:* See Example 1 of this section.)

For a subgroup H of a group G, the sets $x^{-1}Hx$ for each $x \in G$ are called the *conjugates* of H. Exercises 13 through 16 utilize these sets.

13. Show that each conjugate of H is a subgroup of G.

†14. Show that H is isomorphic to each of its conjugates.

15. Prove that H is normal in G if and only if H is its only conjugate.

16. Show that the greatest normal subgroup of G contained in H is the intersection of all the conjugates of H.

4.2   FACTOR GROUPS

We now develop a new group from a given group and a normal subgroup. The new group will turn out to be related to the original group via a homomorphism, thus illustrating another connection between normal subgroups and homomorphisms.

4.2.1 DEFINITION. Let H be a normal subgroup of the group G. We define the *product* of aH and bH to be the coset abH. Symbolically, $(aH) \cdot (bH) = abH$ or, less formally, $aHbH = abH$.

In order to verify that 4.2.1 defines an operation on the set of all left cosets of H in G, we need to show that the product of cosets is independent of the choice of representatives.

4.2.2 THEOREM. Let H be a normal subgroup of a group G. If $xH = aH$ and $yH = bH$, then $xyH = abH$.

Proof: By 3.5.3.3, $x^{-1}a \in H$ and $y^{-1}b \in H$. Let $x^{-1}a = h_1$ and $y^{-1}b = h_2$. H is normal; thus there is an $h_3 \in H$ such that $h_1 b = bh_3$. Now $(xy)^{-1}(ab) = y^{-1}(x^{-1}a)b = y^{-1}h_1 b = y^{-1}bh_3 = h_2 h_3 \in H$ and, again by 3.5.3.3, $xyH = abH.$▲

The normality of H is necessary in the proof of 4.2.2, thus justifying the normality requirement for H in 4.2.1. Exercise 5 of this section shows that, without the normality requirement of H, the product of cosets in 4.2.1 is not an operation on the collection of all left cosets of H in G. Therefore this product is an operation on the left cosets of H in G only when H is normal. Adopting the symbol G/H to represent the collection of all left cosets of H in G when H is normal, we now investigate the basic properties of this operation on G/H.

4.2.3 THEOREM. Let H be a normal subgroup of the group G. Then G/H is a group relative to coset multiplication.

Proof: 1. *Associativity.* Let $aH, bH, cH \in G/H$. Then
$$(aHbH)cH = abHcH = \underbrace{(ab)cH = a(bc)H}_{\text{associativity in } G} = aH(bcH) = aH(bHcH).$$

2. *Identity.* Let $aH \in G/H$. Then $aHeH = aeH = aH$ and $eHaH = eaH = aH$, where e is the identity in G. Therefore $eH = H$ is a left and right identity in G/H and thus is the unique identity.

3. *Inverses.* Let $aH \in G/H$. Since $a \in G$, $a^{-1} \in G$. Now $aHa^{-1}H = aa^{-1}H = eH = H$ and $a^{-1}HaH = a^{-1}aH = eH = H$; thus $(aH)^{-1} = a^{-1}H$.

By (1), (2), and (3), G/H is a group.▲

4.2.4 DEFINITION. The collection of all left cosets of a normal subgroup H of a group G, together with the operation defined in 4.2.1, is called the *factor group of G modulo H* and is denoted G/H.

Recall that $[G:H]$ is the number of left cosets of H in G, so

$|G/H| = [G:H]$; and if $|G|$ is finite, then from Lagrange's theorem, $|G/H| = |G|/|H|$.

EXAMPLE. Let $G = \{e,a,a^2,a^3,a^4,a^5\}$ be the cyclic group of order 6. $H = \{e,a^2,a^4\}$ is a normal subgroup of G. Now $|G/H|$ $= |G|/|H| = 6/3 = 2$ and $G/H = \{H,aH\}$. Also, $N = \{e,a^3\}$ is a normal subgroup of G, $|G/N| = 6/2 = 3$, and $G/N = \{N,aN,a^2N\}$.

There is a natural correspondence between the subgroups of a group G that contain a normal subgroup N and the subgroups of the factor group G/N. We now examine this correspondence.

4.2.5 THEOREM. If N is a normal subgroup of the Group G, then there is a 1–1 correspondence between the subgroups of G/N and the subgroups of G that contain N.

Proof: First consider a subgroup T of G with the property $N \subseteq T$. Since N is normal in G, N is normal in T and the factor group T/N exists. Clearly T/N is a subgroup of G/N.

Now consider a subgroup S of G/N. S is a collection of cosets of N, each of which is a subset of G. Let $B = \{g \in G \mid gN \in S\}$. The identity $e \in B$, so $B \neq \emptyset$. Let $a,b \in B$; then $aN,bN \in S$ and $ab^{-1}N$ $= aN(bN)^{-1} \in S$. This result implies $ab^{-1} \in B$, so B is a subgroup of G. Clearly $N \subseteq B$. Now $T = B$ if and only if $T/N = S$; thus the correspondence is 1–1.▲

4.2.6 THEOREM. If N is a normal subgroup of the group G, then there is a 1–1 correspondence between the normal subgroups of G/N and the normal subgroups of G that contain N.

Proof: Let T be a normal subgroup of G containing N. By 4.2.5, T/N is the unique corresponding subgroup in G/N. We need to show T/N normal in G/N. Let $gN \in G/N$ and $tN \in T/N$; then $(gN)^{-1}tNgN$ $= g^{-1}NtNgN = g^{-1}tgN$. Now $g^{-1}tg \in T$ since T is normal in G. Thus $g^{-1}tgN \in T/N$ and T/N is normal in G/N (4.2.1).

Conversely, let S be a normal subgroup of G/N. By 4.2.5, there is a unique corresponding subgroup T in G such that $N \subseteq T$ and S $= T/N$. We need to show that T is normal in G. Let $g \in G$ and $t \in T$.

Then $g^{-1}tgN = g^{-1}NtNgN \in T/N$ since T/N is normal in G/N. Thus $g^{-1}tg \in T$ and T is normal in G.▲

As an aid to remembering 4.2.5 and 4.2.6, we include the diagram shown in 4.2.7.

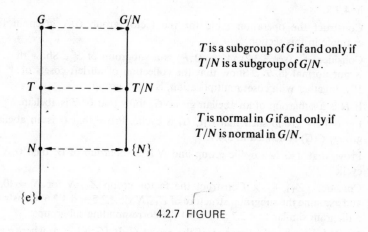

T is a subgroup of G if and only if T/N is a subgroup of G/N.

T is normal in G if and only if T/N is normal in G/N.

4.2.7 FIGURE

The 1–1 correspondences in 4.2.5 and 4.2.6 illustrate the similarity of the subgroup structure of G and G/N only to the extent of the subgroups of G that contain N. These results say nothing about subgroups of G that do not contain N.

We conclude this section by establishing that factor groups are related to the original group via a homomorphism. This result is a converse to 4.1.3 in the sense that it exhibits each normal subgroup of G as the kernel of some homomorphism of G.

4.2.8 THEOREM. Let H be a normal subgroup of the group G. The mapping (called the *natural map*) $v: G \to G/H$ defined by $gv = gH$ for all $g \in G$ is an onto homomorphism and ker $v = H$.

Proof: Clearly v is onto. Let $g_1, g_2 \in G$, then $(g_1 g_2)v = g_1 g_2 H = g_1 H g_2 H = g_1 v g_2 v$ and v is a homomorphism. The identity in G/H is H. Thus we have ker $v = \{g \in G \,|\, gv = H\} = \{g \in G \,|\, gH = H\}$. By Exercise 6 in 3.5, ker $v = H$.▲

This result shows that every factor group of G is a homomorphic image of G. Section 4.3 utilizes this knowledge.

EXERCISES

1. Construct the operation table for the factor group G/K, where G and K are given in 3.5.2.

2. Exhibit the natural homomorphism in Exercise 1 from G onto G/K.

3. If v is the natural homomorphism from G onto G/N, show that $Tv = T/N$ in 4.2.7.

†4. Construct the operation table for the factor group G/N given in the example in this section.

5. Consider $\mathscr{S}_3$ (see 3.3.2). $\mathscr{N} = \{1,P\}$ is a subgroup of $\mathscr{S}_3$. Show that $\mathscr{N}$ is not normal in $\mathscr{S}_3$. Show that the collection of all left cosets of $\mathscr{N}$ in $\mathscr{S}_3$, together with coset multiplication, is not a group.

†6. If H is a subgroup of an abelian group G, show that G/H is abelian.

7. Let G be a group such that $G/C(G)$ is cyclic. Prove that G is an abelian group. $C(G)$ is defined in 3.4.7.

†8. Prove that if G is a cyclic group and N is a subgroup of G, then G/N is cyclic.

9. Consider $\langle Z_{12}, +_{12} \rangle$. Construct the factor group Z_{12}/N for $N = \{\bar{0}, \bar{6}\}$ and examine the subgroup structure of Z_{12}/N via 4.2.5 and 4.2.6. Construct a diagram similar to 4.2.7 and label the corresponding subgroups.

10. Let N be a normal subgroup of the group G. If $[G:N] = p$, where p is a prime, prove that G/N is a simple group.

†11. A proper subgroup M of a group G is a *maximal subgroup* of G if the only subgroup of G that properly contains M is G itself. Prove that a proper normal subgroup with prime index is maximal. (*Hint:* Consider a diagram similar to 4.2.7.)

12. Give an example of a group with at least two maximal subgroups (see Exercise 11).

4.3 THE FUNDAMENTAL THEOREM OF GROUP HOMOMORPHISMS

This section is devoted to establishing a correspondence between factor groups of a group G and general homomorphic images of G. There is a 1–1 correspondence between normal subgroups of G and factor groups of G, since $G/K = G/N$ if and only if $K = N$ (see Exercise 1 of this section). Hence any correspondence between factor groups and general homomorphic images gives rise to a correspondence between normal subgroups of G and general homomorphic images of G.

By 3.2.6, "is isomorphic to" is an equivalence relation on a nonempty collection of groups. The collection $\mathscr{G}$ of all homomorphic images of G is nonempty, since $G \in \mathscr{G}$ (why?). Therefore $\approx$ is an equivalence relation on $\mathscr{G}$ and, by 2.2.6, induces a partition of $\mathscr{G}$ via the equivalence classes. We define two homomorphic images of G to be the same *up to isomorphism* if they belong to the same equivalence class in this partition. Thus if G' and G'' are homomorphic images of G and $G' \approx G''$, then we say that G' and G'' are the same up to isomorphism. The next result establishes the fact that each of these equivalence classes contains at least one factor group—that is, every homomorphic image is a factor group up to isomorphism.

4.3.1 THEOREM. If $\alpha: G \to G'$ is an onto group homomorphism and ker $\alpha = K$, then $G/K \approx G'$.

Proof: By 4.2.5, G/K is a homomorphic image of G via the natural map v, which maps g to gK for each $g \in G$. We define $\beta: G/K \to G'$ by $(gK)\beta = g\alpha$ for each $gK \in G/K$. Exercise 2 of this section verifies that β is a mapping. The diagram shown pictorially represents the information that we have.

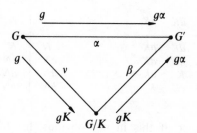

We need to show that β is an isomorphism. To show that the mapping is 1–1, suppose that $(g_1 K)\beta = (g_2 K)\beta$. Then $g_1\alpha = g_2\alpha$, which implies $(g_1\alpha)^{-1}g_2\alpha = e'$, the identity in G'. Now $(g_1^{-1}g_2)\alpha = e'$. Thus $g_1^{-1}g_2 \in K$ and $g_1^{-1}g_2 K = K$; that is, $g_1 K = g_2 K$. For the onto property, let $g' \in G'$. Since α is onto there is a $g \in G$ such that $g\alpha = g'$. Hence $gK \in G/K$ and $(gK)\beta = g\alpha = g'$.

Finally, β has the homomorphism property, since $((g_1 K)(g_2 K))\beta = (g_1 g_2 K)\beta = (g_1 g_2)\alpha = (g_1\alpha)(g_2\alpha) = (g_1 K)\beta(g_2 K)\beta.\blacktriangle$

Now that we have at least one factor group in each equivalence class, it is reasonable to ask if we have exactly one. The following example shows that this is not the case.

EXAMPLE. Let G be the group whose operation is defined by the table in 4.3.2. The group G is abelian; hence all subgroups are

4.3.2 TABLE

·	e	a	b	c	d	f	g	h
e	e	a	b	c	d	f	g	h
a	a	e	d	f	b	c	h	g
b	b	d	e	g	a	h	c	f
c	c	f	g	e	h	a	b	d
d	d	b	a	h	e	g	f	c
f	f	c	h	a	g	e	d	b
g	g	h	c	b	f	d	e	a
h	h	g	f	d	c	b	a	e

normal. Consider the subgroups $K = \{e,a\}$ and $N = \{e,b\}$. The operation tables for G/K and G/N are given in 4.3.3. Note that the groups G/K

4.3.3 TABLE

·	K	cK	dK	hK
K	K	cK	dK	hK
cK	cK	K	hK	dK
dK	dK	hK	K	cK
hK	hK	dK	cK	K

·	N	cN	dN	hN
N	N	cN	dN	hN
cN	cN	N	hN	dN
dN	dN	hN	N	cN
hN	hN	dN	cN	N

and G/N are distinct (if this in not obvious, list the elements of the cosets). Clearly $G/K \approx G/N$.

Theorems 4.1.3, 4.2.8, and 4.3.1 generally are each regarded as part of a major result that gives the complete connection between normal subgroups and group homomorphisms.

4.3.4 THE FUNDAMENTAL THEOREM OF GROUP HOMOMORPHISMS. Associated with each homomorphism of a group G is a normal subgroup of G and associated with each normal subgroup of G is a homomorphism of G having that normal subgroup as its kernel. Furthermore, the only homomorphic images of G are the factor groups of G (up to isomorphism).

EXERCISES

1. Let K and N be normal subgroups of a group G. Prove that $G/K = G/N$ if and only if $K = N$.

2. Let β be defined as in the proof of 4.3.1. Show that β is a mapping by establishing that $aK = bK$ implies $(aK)\beta = (bK)\beta$.

3. In the diagram of 4.3.1, either path from G to G' takes g to $g\alpha$. This is the case since $\alpha = \nu\beta$. Establish that either path from G to G/K takes g to gK by showing that $\nu = \alpha\beta^{-1}$.

†4. Exhibit all homomorphic images (up to isomorphism) of $\langle Z_{12}, +_{12} \rangle$.

†5. List all homomorphic images of a simple group G (up to isomorphism).

6. What can you say about a homomorphic image of a group of prime order? What can you say about the homomorphisms?

7. Let G and G' be groups. Verify that the Cartesian product $G \times G'$ with the operation defined by $(g_1, g'_1)(g_2, g'_2) = (g_1 g_2, g'_1 g'_2)$ is a group. This group is called the *direct product* of G and G'.

†8. (From Exercise 7) Show that G is isomorphic to a normal subgroup of $G \times G'$. (*Hint:* Consider the set $\{(g, e') \mid g \in G\}$.)

9. Let H and K be normal subgroups of a group G. Show that $G/(H \cap K)$ is isomorphic to a subgroup of $G/H \times G/K$. (*Hint:* Consider $\eta : G \to G/H \times G/K$ defined by $g\eta = (gH, gK)$ for all $g \in G$. Show that η is a homomorphism. Find ker η and supply 4.3.1.)

10. If $\alpha : G \to G'$ and $\beta : G \to G''$ are onto group homomorphisms with ker $\alpha =$ ker β, show that G' and G'' are the same up to isomorphism. (*Hint:* Use 4.3.1 twice.)

4.4   THE ISOMORPHISM THEOREMS

We now develop two frequently used theorems in group theory, the isomorphism theorems. Both play an important part in the development of many areas of group theory. Their power is rapidly recognized in a more advanced course in the theory of groups. The remainder of the text is independent of this section. We have chosen to include these results, however, because of their elegance. In addition, they provide some applications of the Fundamental Theorem of Group Homomorphisms and the first develops further the relations between subgroups, their product, and their intersection.

4.4.1 THE FIRST ISOMORPHISM THEOREM. If H and N are subgroups of a group G with N normal in G, then $H/(H \cap N) \approx HN/N$.

Proof: By Exercise 10 in 4.1, $H \cap N$ is normal in H. Exercise 1 in this section verifies that N is normal in HN. Hence $H/(H \cap N)$ and HN/N are meaningful factor groups. Define $\alpha: H \to HN/N$ by $h\alpha = hN$ for each $h \in H$. Consider the diagram shown. If we can show α is an

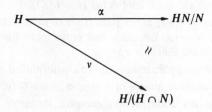

$$H \xrightarrow{\quad \alpha \quad} HN/N$$

$$H/(H \cap N)$$

onto homomorphism with ker $\alpha = H \cap N$, then by 4.3.1 we are finished. First we show α is onto. Let $xN \in HN/N$. Then $x \in HN$, so we have $x = hn$ for suitable $h \in H$ and $n \in N$. Therefore $xN = hnN = hN = h\alpha$.

In order to show that α is a homomorphism, let $h_1, h_2 \in H$. Then $(h_1 h_2)\alpha = h_1 h_2 N = h_1 N h_2 N = (h_1\alpha)(h_2\alpha)$.

Finally, we show that ker $\alpha = H \cap N$. Let $x \in$ ker α; then $x\alpha = xN = N$. By Exercise 6 in 3.5, $x \in N$. Clearly $x \in$ ker α implies $x \in H$. Thus $x \in H \cap N$ and ker $\alpha \subseteq (H \cap N)$. Now let $y \in (H \cap N)$. Then $y \in N$, so $y\alpha = yN = N$. Hence $y \in$ ker α and $(H \cap N) \subseteq$ ker α. Therefore we have equality.▲

4.4.2 THE SECOND ISOMORPHISM THEOREM. Let $\alpha: G \to G'$ be an onto group homomorphism. If H' is a normal subgroup of G' and $H = \{g \in G \mid g\alpha \in H'\}$, then H is a normal subgroup of G and $G/H \approx G'/H'$.

Proof: By Exercise 6 in 3.5, H is a subgroup of G. To show H is normal in G, let $g \in G$ and $h \in H$. Then $g\alpha \in G'$ and $h\alpha \in H'$. Since H' is normal in G', we have $(g^{-1}hg)\alpha = (g\alpha)^{-1}(h\alpha)(g\alpha) \in H'$. Therefore $g^{-1}hg \in H$ and H is normal in G.

Define $\beta: G/H \to G'/H'$ by $(gH)\beta = g\alpha H'$ for all $gH \in G/H$. Clearly since α is onto, β is also onto. To show that β is 1-1, suppose that $(g_1 H)\beta = (g_2 H)\beta$. Then $g_1\alpha H' = g_2\alpha H'$ and $(g_1\alpha)^{-1}(g_2\alpha)H' = H'$; that is, $(g_1^{-1}g_2)\alpha H' = H'$. Therefore $g_1^{-1}g_2 \in H$ and so $g_1 H = g_2 H$.

For the homomorphism property, let $g_1 H, g_2 H \in G/H$. Then $(g_1 H g_2 H)\beta = (g_1 g_2 H)\beta = (g_1 g_2)\alpha H' = (g_1\alpha)(g_2\alpha)H' = (g_1\alpha H')(g_2\alpha H')$ $= (g_1 H)\beta(g_2 H)\beta$ and β is an isomorphism.▲

More frequently used than the Second Isomorphism Theorem is the following corollary, which is easily remembered because it seems to treat factor groups like fractions.

4.4.3 COROLLARY. If H and N are normal subgroups of a group G with $N \subseteq H$, then $G/H \approx (G/N)/(H/N)$.

Proof: Let α be the natural homomorphism from G onto G/N; that is, $g\alpha = gH$ for all $g \in G$. By 4.2.6, since H is normal in G and ker $\alpha = N \subseteq H$, then H/N is normal in G/N. Moreover,
$$H = \{g \in G \mid g\alpha \in H/N\};$$
so, by 4.4.2, $G/H \approx (G/N)/(H/N)$. ▲

EXERCISES

†1. Let H and N be subgroups of G with N normal in G. Prove that N is normal in HN.

2. Prove the following equivalent statement of the Second Isomorphism Theorem. Let $\alpha: G \to G'$ be an onto group homomorphism. If H is a normal subgroup of G and ker $\alpha \subseteq H$, then $H' = H\alpha$ is a normal subgroup of G' and $G/H \approx G'/H'$.

3. A normal subgroup N of a group G is a *maximal normal subgroup* of G if $N \subset G$ and there exists no normal subgroup K of G with $N \subset K \subset G$. Prove that H is a maximal normal subgroup of G if and only if G/H is simple. (*Hint:* Use Exercise 2.)

4. Prove that if H and N are distinct maximal normal subgroups of a group G (see Exercise 3), then $G/H \approx N/(H \cap N)$ and $G/N \approx H/(H \cap N)$. (*Hint:* Show that $G = HN = NH$ and use the First Isomorphism Theorem.)

5. Assume the hypotheses of 4.4.2. Define $\gamma: G \to G'/H'$ by $g\gamma = (g\alpha)H'$ for all $g \in G$. Show that γ is an onto homomorphism with ker $\gamma = H$. Then apply 4.3.4 to complete a more elegant proof of the Second Isomorphism Theorem.

6. Let H and N be normal subgroups of a group G with $N \subseteq H$. Define $\alpha: G \to (G/N)/(H/N)$ by $g\alpha = (gN)H/N$ for all $g \in G$. Show that α is an onto homomorphism with ker $\alpha = H$. Then apply 4.3.4 to conclude that $G/H \approx (G/N)/(H/N)$. This is an alternate proof for 4.4.3.

†7. Let H and N be subgroups of a group G with N normal in G. Prove that HN is the smallest subgroup of G that contains $H \cup N$ (smallest in the sense that any subgroup of G containing $H \cup N$ also contains HN).

8. Illustrate the First Isomorphism Theorem when $H = \langle 2 \rangle$ and $N = \langle 3 \rangle$ are considered as subgroups of $\langle Z, + \rangle$.

4.5   ADDITIVE NOTATION—RING THEORY REFERENCE

It was notationally convenient to denote the group operation multiplicatively while developing the theory of groups. The next algebraic system we study has two operations that are usually denoted by + and juxtaposition. The additive structure, if considered by itself, yields an abelian group. Therefore we are faced with the problem of applying material that was originally developed in a "multiplicatively oriented environment" in this new "additive environment." In this section the more relevant definitions and theorems are restated in additive notation. They will keep their original identification numbers as an aid to the location of their multiplicative counterpart and proof. We simply affix an "a" to the number to emphasize that it is a restatement of previously studied material. For example, 3.1.1 in additive notation will be identified by 3.1.1a. We hope this section facilitates the use of group theory in the study of rings.

3.1.1a DEFINITION. A *group* is an algebraic system consisting of a nonempty set G and an operation + on G having the following properties:

1. + is associative in G; that is, for all $a,b,c \in G$, $(a + b) + c = a + (b + c)$.

2. G has an identity relative to +; that is, there exists an element $0 \in G$ such that $a + 0 = 0 + a = a$ for all $a \in G$.

3. Each element of G has an inverse; that is, for each $g \in G$ there exists $-g \in G$ such that $g + (-g) = (-g) + g = 0$.

3.1.3a THEOREM. The left and right cancellation laws hold in $\langle G, + \rangle$; that is, for $a,b,c \in G$, if $a + b = a + c$ or $b + a = c + a$, then $b = c$.

3.1.5a COROLLARY. In $\langle G, + \rangle$, the inverse of a "sum" is the "sum" of the inverses in the reverse order; that is, for $a,b \in G$, $-(a + b) = (-b) + (-a)$.

3.1.6a COROLLARY. In $\langle G, + \rangle$, the inverse of the inverse of an element in G is the element; that is, for $g \in G$, $-(-g) = g$.

3.2.1a DEFINITION. $\alpha: G_1 \to G_2$ is a *group homomorphism* from $\langle G_1,+\rangle$ to $\langle G_2,\oplus\rangle$ if for all $a,b \in G_1$ we have $(a + b)\alpha = a\alpha \oplus b\alpha$.

3.2.3a THEOREM. If α is a homomorphism from $\langle G_1,+\rangle$ to $\langle G_2,\oplus\rangle$, then
1. if the respective identities are 0_1 and 0_2, then $0_1\alpha = 0_2$ and
2. for $g \in G_1$, $(-g)\alpha = -(g\alpha) \in G_2$.

3.4.3a THEOREM. Let H be a nonempty subset of the group G. The following statements are equivalent.
1. H is a subgroup of G.
2. For all $a,b \in H$ we have $(a + b) \in H$ and $-a \in H$.
3. For all $a,b \in H$ we have $[a + (-b)] \in H$.

3.5.1a DEFINITION. Let G be a group, H a subgroup of G, and $g \in G$.
1. $g + H = \{g + h \mid h \in H\}$ is called the *left coset* of g and H in G.
2. $H + g = \{h + g \mid h \in H\}$ is called the *right coset* of g and H in G.

3.5.3a THEOREM. Let G be a group H a subgroup of G and $a,b \in G$. The following statements are equivalent.
1. a and b belong to the same left coset of H in G (i.e., $a \in b + H$).
2. a and b determine the same left coset of H in G (i.e., $a + H = b + H$).
3. $[(-a) + b] \in H$.

3.7.1a DEFINITION. Let $g \in G$, where G is a group with identity $0'$.
1. $0g = 0'$. (Note that 0 is the integer zero while $0'$ is the additive identity in G.)
2. $mg = g + g + \cdots + g$ for a positive integer m.
3. $(-m)g = m(-g)$ for each positive integer m. (Note that $-m$ is a negative integer while $-g$ is the additive inverse of $g \in G$.)

3.7.2a LEMMA. Let $m \in Z_+$ and g be an element of the group G. Then

1. $-(mg) = m(-g)$, and

2. negative one times g is the additive inverse of g; that is, $(-1)g = -g$.

3.7.3a THEOREM. Let G be a group, $g \in G$, and $m \ n, \in Z$. Then

1. $(mg) + (ng) = (m + n)g$. (Note that the first plus sign represents the operation in G, whereas the second plus sign represents addition in Z—the two are usually not related!)

2. $n(mg) = (nm)g$. (Note that $n(mg)$ is n addends of mg, whereas $(nm)g$ is nm addends of g.)

3.7.4a DEFINITION. Let g be an element of the group G. The subset of G consisting of all integral multiples of g is called the *cyclic subgroup generated by g*. Symbolically,

$$\langle g \rangle = \{ng \mid n \in Z\}.$$

3.7.8a THEOREM. Let g be an element of finite order in a group G. Then $|g| = n$ if and only if n is the smallest positive integer such that $ng = 0$, where 0 is the additive identity in G.

3.7.10a COROLLARY. Let G be a group with identity 0. If $|G| = n$, then $ng = 0$ for all $g \in G$.

4.1.1a DEFINITION. Let H be a subgroup of the group G. H is called a *normal subgroup* of G if for all $g \in G$ we have $g + H = H + g$.

4.1.2a THEOREM. Let H be a subgroup of the group G. The following statements are equivalent.

1. H is normal in G (i.e., $g + H = H + g$ for all $g \in G$).
2. $g + H \subseteq H + g$ for all $g \in G$.
3. $(-g) + H + g \subseteq H$ for each $g \in G$.
4. $(-g) + H + g = H$ for each $g \in G$.
5. $(-g) + h + g \in H$ for each $g \in G$ and for each $h \in H$.

4.2.1a DEFINITION. Let H be a normal subgroup of the group G. We define the sum of $a + H$ and $b + H$ to be the left coset determined by $a + b$. Symbolically, $(a + H) + (b + H) = (a + b) + H$.

CHAPTER 5

RINGS

In this chapter we will consider an algebraic system that has more of the characteristics and properties of the familiar number systems than the group. Recall that $\langle Z,+\rangle$, $\langle Q,+\rangle$, and $\langle F,+\rangle$ exemplify the group concept. In considering these systems, we ignored the existence of the multiplicative structure (except when we considered $\langle Q_0,\cdot\rangle$ and $\langle F_0,\cdot\rangle$). By considering these systems with both operations present, we are led to what is called the *theory of rings*.

5.1   DEFINITION OF A RING

Our second algebraic system is a generalization of the familiar two-operation number systems and, as such, is required to possess essentially the same basic properties.

5.1.1 DEFINITION. A *ring* $\langle R, +, \cdot \rangle$ is an algebraic system consisting of a nonempty set R and two binary operations, usually denoted by $+$ and $\cdot$ (or juxtaposition), so that:

1. $\langle R, + \rangle$ is an abelian group,
2. $\cdot$ is associative on R, and
3. $\cdot$ is distributive over $+$; that is, $a(b + c) = ab + ac$ and $(a + b)c = ac + bc$ for all $a, b, c \in R$.

It should be noted that we are merely using the symbols $+$ and $\cdot$ to represent the two operations in a ring and that this does not, in general, signify a relationship between these operations and the usual concepts of addition and multiplication of ordinary numbers. Since the "additive" structure in a ring satisfies the group concept, a cursory reading of Section 4.5 is recommended at this time.

In Section 3.1, the uniqueness of the identity and inverses in a group was established. Our definition of a ring requires $\langle R, + \rangle$ to be an abelian group. This guarantees, in $\langle R, +, \cdot \rangle$, the uniqueness of the additive identity, 0, and of the additive inverse, $-a$, for each a in R.

EXAMPLES OF RINGS

1. From the familiar number systems we have $\langle Z, +, \cdot \rangle$, $\langle Q, +, \cdot \rangle$ and $\langle F, +, \cdot \rangle$.

2. $\langle \{0\}, +, \cdot \rangle$.

3. $\langle Z_m, +_m, \cdot_m \rangle$ for each $m \in Z_+$ (see 2.7.7 through 2.7.11).

4. $\langle P(S), +, \cdot \rangle$, where $P(S)$ is the power set of S and addition and multiplication are defined as follows: For $A, B \in P(S)$, $A + B = (A \cup B) \setminus (A \cap B)$ and $A \cdot B = A \cap B$.

5. The set of all 2×2 matrices of real numbers with matrix addition and matrix multiplication; that is, $\left\{ \begin{pmatrix} a\ b \\ c\ d \end{pmatrix} \,\middle|\, a,b,c,d \in F \right\}$ with addition defined by

$$\begin{pmatrix} s & t \\ u & v \end{pmatrix} + \begin{pmatrix} w & x \\ y & z \end{pmatrix} = \begin{pmatrix} s+w & t+x \\ u+y & v+z \end{pmatrix};$$

and multiplication defined as in Example 6 of 3.1.

6. $\langle R,+,\cdot \rangle$, where R is the set of even integers.

7. $\langle R,+,\cdot \rangle$, where $R = \{o,a,b,c\}$ and addition and multiplication are defined by the following tables:

+	o	a	b	c
o	o	a	b	c
a	a	o	c	b
b	b	c	o	a
c	c	b	a	o

$\cdot$	o	a	b	c
o	o	o	o	o
a	o	a	b	c
b	o	o	o	o
c	o	a	b	c

8. The set of 2×2 matrices of even integers

$$\left(\left\{ \begin{pmatrix} a & b \\ c & d \end{pmatrix} \middle| \; a,b,c,d \text{ are even integers} \right\} \right)$$

with addition and multiplication as in Example 5.

Rings are generally divided into classes via two additional properties that the operation of multiplication may or may not possess. These properties are (1) commutativity of multiplication and (2) the existence of a multiplicative identity. The terms commutative ring, noncommutative ring, ring with identity, and so on, are used to label these classes. More formally:

5.1.2 DEFINITION.

1. $\langle R,+,\cdot \rangle$ is a *commutative ring* if $ab = ba$ for all $a,b \in R$.

2. $\langle R,+,\cdot \rangle$ is a *ring with identity* if there exists an $e \in R$ such that $ae = ea = a$ for all $a \in R$.

3. $\langle R,+,\cdot \rangle$ is a *commutative ring with identity* if multiplication is commutative on R and there exists a multiplicative identity.

When these terms are used to categorize rings, they refer only to the multiplicative structure. The additive structure of a ring, as an abelian group, has these properties; thus it would be redundant to mention them in this sense.

The following table classifies the examples in this section according to the above discussion.

	Identity	Nonidentity
Commutative	1, 2, 3, 4	6
Noncommutative	5	7, 8

Noncommutative rings are treated extensively in more advanced ring theory. However, for our purpose, their existence alone forces us to exercise care in proving results for rings. In other words, if commutativity is required in a proof, we must be sure it is hypothesized or proved first.

EXERCISES

1. Let $S = \{a,b\}$. $P(S)$ has four elements. Thus the ring whose elements are subsets of S (see Example 4) has four elements. Make addition and multiplication tables for this ring. What is the identity?

†2. Verify that Example 4 is a commutative ring with identity (see Exercise 11 in Section 3.1).

3. (Refer to Example 5) What is the identity for this ring? Verify that the distributive properties hold.

†4. Which of the following structures are rings with the usual definitions of addition and multiplication?
   (a) The set of numbers of the form $a + b\sqrt{2}$ for $a, b \in Z$.
   (b) The set of all negative integers.
   (c) The set of all irrational numbers—that is, the set of all real numbers that are not rational numbers.
   (d) The set of all integral multiples of 3; that is, $\{3n \mid n \in Z\}$.

5. Prove that the set R of all integral multiples of a fixed integer $k > 1$, together with ordinary addition and multiplication, is a commutative ring without identity ($R = \{nk \mid k > 1$ is a fixed integer and $n \in Z\}$). (*Note:* The ring R reduces to the ring in Example 6 when $k = 2$ and reduces to the ring in Exercise 4(d) when $k = 3$.)

6. Find two elements in the ring R of Example 7 that do not commute (relative to multiplication). This verifies that R is noncommutative.

7. Construct addition and multiplication tables for $\langle Z_6, +_6, \cdot_6 \rangle$.
   (a) Find an example to show that the cancellation law for multiplication does not hold; that is, show by counterexample that $a \neq \bar{0}$ and $ab = ac$ does not imply $b = c$.
   (b) Find an example to show that $ab = \bar{0}$ does not imply $a = \bar{0}$ or $b = \bar{0}$.

8. Let $\langle R, + \rangle$ be an abelian group. Define multiplication as follows: $a \cdot b = 0$ for all $a, b \in R$. Prove that $\langle R, +, \cdot \rangle$ is a commutative ring. Does $\langle R, +, \cdot \rangle$ have an identity?

†9. Define $\oplus$ and $\odot$ as follows:
$$a \oplus b = a + b - 1$$
and $a \odot b = a + b - ab$ for all $a,b \in Z$. Verify that $\langle Z, \oplus, \odot \rangle$ is a commutative ring with identity.

10. Let $R = Z \times Z$. Define $\oplus$ and $\odot$ as follows:
$$(a,b) \oplus (c,d) = (a + c, b + d)$$
and $(a,b) \odot (c,d) = (ac,bd)$ for all $(a,b),(c,d) \in R$. Show that $\langle R,\oplus,\odot \rangle$ is a commutative ring with identity.

11. Let $R = Z \times Z$. Define $\oplus$ and $\odot$ as follows:
$$(a,b) \oplus (c,d) = (a + c, b + d)$$
and $(a,b) \odot (c,d) = (ac - bd, ad + bc)$ for all $(a,b),(c,d) \in R$. Show that $\langle R,\oplus,\odot \rangle$ is a commutative ring with identity. $\langle R,\oplus,\odot \rangle$ is called the *Ring of Gaussian Integers*.

5.2   ELEMENTARY PROPERTIES

In 2.5.3 we established that the existence of an identity for an operation on a set implied uniqueness of the identity. Since multiplication is an operation in a ring, we conclude that when an identity exists it is unique. Similarly, by 2.5.6, multiplicative inverses are unique when they exist.

In an arbitrary ring, we have no guarantee of the existence of an identity or of multiplicative inverses. For instance, in Examples 6, 7, and 8 in Section 5.1, there is no identity and hence no multiplicative inverses. In the ring of integers, a commutative ring with identity, the only elements that have multiplicative inverses are 1 and -1. At the other extreme are rings like $\langle F,+,\cdot \rangle$ and $\langle Z_7,+_7,\cdot_7 \rangle$ in which every nonzero element has a multiplicative inverse. Elements having inverses with respect to multiplication will be studied in Section 6.2.

Additional elementary properties of the familiar number systems remain valid in the more general structure of a ring. For example, the property that "any number times zero equals zero" holds in every ring.

5.2.1 THEOREM. In a ring $\langle R,+,\cdot \rangle$ with additive identity denoted by 0, we have $r0 = 0r = 0$ for all $r \in R$.

Proof: Let $r \in R$.
$$r0 = r(0 + 0)$$
$$= r0 + r0. \qquad \text{(Why?)}$$

Now $r0 + 0 = r0 + r0$ and, by 3.1.3a in Section 4.5, we have $0 = r0$. Similarly, $0r = 0$.▲

Another familiar property is exemplified in the system of integers by $5(-4) = (-5)(4) = -(5 \cdot 4)$ and still another by the phrase "A minus times a minus is a plus." These properties also hold in the general ring.

As in the study of groups, we now drop the formality of writing $\langle R,+,\cdot \rangle$ for a ring except when different symbols are used for the operations. A single letter will be used to denote a ring and, unless otherwise stated, $+$ and $\cdot$ (or juxtaposition) will denote the operations.

5.2.2 THEOREM. Let R be a ring and $a,b \in R$.
Then

      1. $a(-b) = -(ab)$,
      2. $(-a)b = -(ab)$,
      3. $(-a)(-b) = ab$.

Proof: 1. $-(ab)$ is the unique additive inverse of ab. Now
$$ab + a(-b) = a[b + (-b)] \quad \text{(Why?)}$$
$$= a(0)$$
$$= 0.$$
Thus $a(-b)$ is also the unique additive inverse of ab and we must have $a(-b) = -(ab)$.

      2. Exercise 1 of this section.
      3. $(-a)(-b) = -[(-a)b]$ by (1)
$$= -[-(ab)] \quad \text{by (2)}$$
$$= ab \quad \text{by 3.1.6a in 4.5.}▲$$

Subtraction, as an operation in the number systems, is defined in terms of addition. For example, $3 - 5$ is defined to be the number $3 + (-5)$. The "minus" sign in $3 - 5$ refers to the operation of subtraction and the "minus" sign in $3 + (-5)$ refers to the additive inverse of 5. We retain the same definition and symbolism in ring theory.

5.2.3 DEFINITION. Let R be a ring. *Subtraction* is defined as follows:
$$a - b = a + (-b) \quad \text{for all } a,b \in R.$$

As a consequence of this definition, we have the following immediate results.

5.2.4 THEOREM. Let R be a ring and $a,b,c \in R$. Then
1. $-(a + b) = -a - b$,
2. $-(a - b) = b - a$,
3. $(a - b) - c = (a - c) - b$,
4. $a(b - c) = ab - ac$,
5. $(a - b)c = ac - bc$.

Proof: 4. $a(b - c) = a[b + (-c)]$ (Why?)
$$= ab + a(-c)$$
$$= ab + [-(ac)]$$
$$= ab - ac. \quad \text{(Why?)}$$
Results (1), (2), (3), and (5) are left for Exercise 2 of this section. ▲

EXERCISES

1. Prove part (2) in 5.2.2.
2. Complete the proof of 5.2.4.
†3. Prove 5.2.1 by using the fact that $\langle R, + \rangle$ is a group and the only additive idempotent (see Exercise 10 of 3.1) element is the additive identity.
4. Determine the elements in $\langle Z_6, +_6, \cdot_6 \rangle$ that have multiplicative inverses.
†5. Let R be a ring with identity e. Show that $(-e)(-e) = e$.
6. Let R be a ring and $a, b \in R$. Prove that $(a - b)(a + b) = a^2 - ba + ab - b^2$. What requirement must be imposed on R if
$$(a - b)(a + b) = a^2 - b^2$$
for all $a, b \in R^2$?
7. Show that $(-a)(-b)(-c) = -abc$ for all elements $a, b,$ and c in a ring.
†8. Let R be a ring in which $a^2 = a$ for all $a \in R$. Prove:
(a) $a + a = 0$ for all $a \in R$, and (b) R is commutative.
(*Hint:* $(a + a)^2 = a + a$ and $(a + b)^2 = a + b$).

5.3 RING HOMOMORPHISM AND RING ISOMORPHISM

The same basic techniques used to obtain information about groups are employed in the study of rings; namely,

1. a study of mappings between rings that preserve the corresponding ring operations,
2. a study of some particular subsets of a ring, and
3. a study of ring decompositions via equivalence relations.

We proceed with the first of these.

Mappings that preserve ring operations are slightly more sophisticated than mappings that preserve a group operation simply because there is more to preserve in a ring. Two operations must be simultaneously preserved—that is, in rings R and R' we want the image of a sum of elements in R to be the sum of the corresponding images in R' and, also, the image of a product of elements in R to be the product of the corresponding images in R'. More formally:

5.3.1 DEFINITION. $\alpha: R \to R'$ is a *ring homomorphism* from $\langle R, +, \cdot \rangle$ to $\langle R', \oplus, \odot \rangle$ if α has the following properties: for all $a,b \in R$
   1. $(a + b)\alpha = a\alpha \oplus b\alpha$, and
   2. $(ab)\alpha = a\alpha \odot b\alpha$.

While (1) and (2) are expressions between elements in R', it is important to consider them in the following manner. The left sides of the expressions say "operate in R and then map" and the right sides say "map first into R' and then operate on the images." The equality states that both methods produce the same result.

As in group theory, the range of α, $R\alpha$, is called a *homomorphic image* of R, and if α is onto ($R\alpha = R'$), then R and R' are said to be *homomorphic rings*. Schematically, a ring homomorphism can be pictured as shown in the diagram.

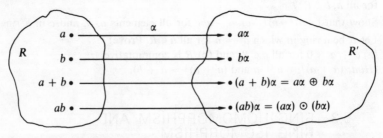

If we restrict our attention to the additive structures of R and R', then α can be considered as a group homomorphism. Therefore all our results from group theory concerning group homomorphisms hold for the additive structure of rings. In particular, $0\alpha = 0'$ (0 and $0'$ are the additive identities of R and R' respectively) and $(-a)\alpha = -(a\alpha)$ for all $a \in R$. Thus not just any mapping from R to R' can be a ring homomorphism.

The next result, analogous to 3.2.4, shows that we could adopt the convenience of considering only onto ring homomorphisms without any loss of generality.

5.3.2 THEOREM. If α is a ring homomorphism from $\langle R, +, \cdot \rangle$ to $\langle R', \oplus, \odot \rangle$, then $R\alpha$, the homomorphic image of R, together with $\oplus$ and $\odot$, is a ring.

Proof: By 3.2.4 and Exercise 3 in Section 3.2, $\langle R\alpha, \oplus \rangle$ is an abelian group. To establish that $\odot$ is an operation on $R\alpha$ let $x, y \in R\alpha$. Then there exist $a, b \in R$, so that $a\alpha = x$ and $b\alpha = y$. Now $x \odot y = a\alpha \odot b\alpha = (ab)\alpha$ and $ab \in R$. Thus $(x \odot y) \in R\alpha$.

Finally, associativity for $\odot$ and distributivity of $\odot$ over $\oplus$ are inherited from R'. Therefore $\langle R\alpha, \oplus, \odot \rangle$ is a ring.▲

EXAMPLES

1. $\alpha : R \to R'$ defined by $a\alpha = 0'$ ($0'$ is the additive identity in R') for all $a \in R$ is a trivial ring homomorphism from the ring R to the ring R'.

2. $\eta : Z_8 \to Z_4$ defined by the following is a ring homomorphism from $\langle Z_8, +_8, \cdot_8 \rangle$ onto $\langle Z_4, +_4, \cdot_4 \rangle$. For each $\bar{x} \in Z_8$, $\bar{x}\eta = \bar{r}$, where $r \overset{\sim}{4} x$ and $\bar{r} \in Z_4$ (see Example 4 in Section 3.2). As one instance of the homomorphism property for multiplication,
$$(\bar{3} \cdot_8 \bar{5})\eta = \bar{7}\eta = \bar{3} \quad \text{and} \quad \bar{3}\eta \cdot_4 \bar{5}\eta = \bar{3} \cdot_4 \bar{1} = \bar{3}.$$
Thus
$$(\bar{3} \cdot_8 \bar{5})\eta = (\bar{3}\eta) \cdot_4 (\bar{5}\eta).$$
In general,
$$(\bar{a} \cdot_8 \bar{b})\eta = (\bar{a}\eta) \cdot_4 (\bar{b}\eta) \qquad \text{for all } \bar{a}, \bar{b} \in Z_8.$$
Is $\bar{1}$ the identity for Z_8? What is the image of $\bar{1}$ under η? Does Z_4 have an identity? If so, is it $\bar{1}\eta$?

3. $\beta : Z \to Z_2$ defined by
$$a\beta = \begin{cases} \bar{0} \text{ if } a \text{ is even} \\ \bar{1} \text{ if } a \text{ is odd} \end{cases} \qquad \text{for all } a \in Z$$
is a ring homomorphism from Z onto Z_2.

Ring homomorphisms, in addition to preserving ring operations, also preserve the properties that we use to differentiate between various types of rings.

5.3.3 THEOREM.
1. The homomorphic image of a commutative ring is a commutative ring.
2. The homomorphic image of a ring with identity is a ring with identity.

Proof: Exercise 2 of this section.▲

5.3.4 COROLLARY. The homomorphic image of a commutative ring with identity is a commutative ring with identity.

Proof: The result is an immediate consequence of 5.3.3.▲

We now consider the special case when a ring homomorphism is a 1–1 correspondence. In this instance, as with isomorphic groups, we may consider the rings to differ only in the notation used for the elements and the operations. If α is a 1–1 ring homomorphism from R_1 onto R_2, then $\langle R_2, \oplus, \odot \rangle$ is simply $\langle R_1, +, \cdot \rangle$ in disguise.

5.3.5 DEFINITION. A 1–1 ring homomorphism from R_1 onto R_2 is a *ring isomorphism* from R_1 onto R_2. We write $R_1 \approx R_2$ and say R_1 and R_2 are *isomorphic rings*.

If $\alpha: R_1 \to R_2$ is a ring isomorphism, then $\alpha^{-1}: R_2 \to R_1$ is also a ring isomorphism. Furthermore, the relation $\approx$ on a collection of rings is an equivalence relation (see 3.2.6).

EXAMPLES

1. Let R be a ring. i_R, the identity map on R, is a ring isomorphism.
2. Let R denote the ring given in Exercise 9 in Section 5.1. Then $R \approx Z$, since $\alpha: Z \to R$, defined by $a\alpha = 1 - a$ for all $a \in Z$, is a ring isomorphism. Clearly α is 1–1 and onto. In order to see that α has the homomorphism properties, let $a, b \in Z$. Then

$$(a + b)\alpha = 1 - (a + b)$$
$$= (1 - a) + (1 - b) - 1$$
$$= (1 - a) \oplus (1 - b)$$
$$= a\alpha \oplus b\alpha,$$
$$(ab)\alpha = 1 - ab$$
$$= (1 - a) + (1 - b) - (1 - a)(1 - b)$$
$$= (1 - a) \odot (1 - b)$$
$$= a\alpha \odot b\alpha.$$

EXERCISES

1. Verify that β in Example 3 is a ring homomorphism.

2. Prove 5.3.3.

†3. Define $\alpha: Z \to Z$ by $a\alpha = 2a$ for all $a \in Z$. Why is α not a ring isomorphism from the ring of integers onto itself?

4. $\bar{1}, \bar{3}, \bar{5}, \bar{7} \in Z_8$ all have multiplicative inverses. Do their images under η (see Example 2) have multiplicative inverses?

†5. Let $\alpha: R_1 \to R_2$ be an onto ring homomorphism and let R_1 have identity e. If $a \in R_1$ has an inverse (multiplicative), prove that $a\alpha \in R_2$ has an inverse.

6. Let R denote the ring in Exercise 10 of Section 5.1. Define $\alpha: R \to Z$ by $(a,b)\alpha = a$ for all $(a,b) \in R$. Show that α is a ring homomorphism from R onto Z.

7. If R denotes the ring in Exercise 11 of Section 5.1, why is $\alpha: R \to Z$, defined by $(a,b)\alpha = a$ for all $(a,b) \in R$, not a ring homomorphism from R onto the ring of integers?

8. $R = \{a + b\sqrt{2} \,|\, a,b \in Z\}$ with the usual operations of addition and multiplication is a ring by Exercise 4(a) in Section 5.1. Show that $\eta: R \to R$ defined by $(a + b\sqrt{2})\eta = a - b\sqrt{2}$ for all $a,b \in Z$ is a ring isomorphism.

†9. Give an example to show that the homomorphic image of a noncommutative ring might be commutative. (*Hint:* Construct a homomorphism from R onto Z_2 with R denoting the ring of Example 7 in Section 5.1. This example also shows that a ring without identity might have a homomorphic image with identity.)

†10. Let Aut (R) denote the set of all ring isomorphisms on R. Each element of Aut (R) is called a *ring automorphism*.

    (a) Prove that Aut (R) is a group with respect to composition.

    (b) Let $S \subseteq R$. Show that $K = \{\alpha/\alpha \in$ Aut (R) and $a\alpha = a$ for all $a \in S\}$ is a subgroup of Aut (R).

11. Let $\langle R, +, \cdot \rangle$ be a commutative ring with identity e. Define $\oplus$ and $\odot$ as follows: For all $a,b \in R$,

$$a \oplus b = a + b - e,$$

and

$$a \odot b = a + b - ab.$$

Show that $\langle R, \oplus, \odot \rangle$ is a commutative ring with identity; then prove that $\langle R, +, \cdot \rangle \approx \langle R, \oplus, \odot \rangle$. (*Hint:* Construct a mapping similar to α in Example 2, p.106.)

5.4   RINGS OF MAPPINGS

Cayley's theorem (3.3.4) states that any group can be considered as a group of mappings (permutations). We will develop a similar result for rings. This section contains the result for rings with identity and Section 5.7 contains the more general result.

The mappings considered in obtaining Cayley's theorem were quite special—namely, 1–1 correspondences between a set and itself. It is not surprising, therefore, that the mappings we consider are also quite special. Yet they are not 1–1 correspondences as the permutations were. Each of the mappings that we will construct is a group homomorphism from an abelian group to itself. The proof of this result is divided into several parts so that, hopefully, both the "forest" and the "trees" are visible.

5.4.1 THEOREM. Let R be a ring with identity e. Then R is isomorphic to a ring of group homomorphisms from an abelian group to itself.

Proof: The mappings. We first construct the desired group homomorphisms on the only abelian group in sight, namely, $\langle R, + \rangle$. For $a \in R$, define $\alpha_a : R \to R$ by $x\alpha_a = xa$ for each $x \in R$. Consider $\mathscr{R} = \{\alpha_a | a \in R\}$ and let $\alpha_a \in \mathscr{R}$. Then for all $x, y \in R$, we have

$$(x + y)\,\alpha_a = (x + y)a$$
$$= xa + ya$$
$$= x\alpha_a + y\alpha_a.$$

Therefore $\mathscr{R}$ is a collection of group homomorphisms from $\langle R, + \rangle$ to $\langle R, + \rangle$.

The ring of mappings. We now have a collection, $\mathscr{R}$, of group homomorphisms and need two operations, $\oplus$ and $\odot$, on $\mathscr{R}$ that have the required ring properties. The operations on R are used to induce the corresponding operations on $\mathscr{R}$. For $\alpha_a, \alpha_b \in \mathscr{R}$, we define and

$$\alpha_a \oplus \alpha_b = \alpha_{a+b}$$

$$\alpha_a \odot \alpha_b = \alpha_{ab}.$$

Exercise 1 in this section verifies that $\mathscr{R}$, together with $\oplus$ and $\odot$, is a ring.

The isomorphism. Define $\beta: R \to \mathscr{R}$ in the natural way by $a\beta = \alpha_a$ for each $a \in R$. β has the homomorphism properties, since for each $a,b \in R$ we have

$$(a + b)\beta = \alpha_{a+b} \qquad (ab)\beta = \alpha_{ab}$$
$$= \alpha_a \oplus \alpha_b \qquad\qquad = \alpha_a \odot \alpha_b$$
$$= a\beta \oplus b\beta, \qquad\qquad = a\beta \odot b\beta.$$

Clearly β is onto. To show that β is 1–1, let $a\beta = b\beta$. Then $\alpha_a = \alpha_b$, and for all $x \in R$, $x\alpha_a = x\alpha_b$. In particular, $e\alpha_a = e\alpha_b$; that is, $ea = eb$ and hence $a = b$.

Finally, β is an isomorphism and we conclude that $R \approx \mathscr{R}$. ▲

EXAMPLE. To clarify the proof of 5.4.1, take $R = Z_3$. Then $\mathscr{R} = \{\alpha_{\bar{0}}, \alpha_{\bar{1}}, \alpha_{\bar{2}}\}$ and $\oplus$ and $\odot$ are as shown in the tables.

$\oplus$	$\alpha_{\bar{0}}$	$\alpha_{\bar{1}}$	$\alpha_{\bar{2}}$
$\alpha_{\bar{0}}$	$\alpha_{\bar{0}}$	$\alpha_{\bar{1}}$	$\alpha_{\bar{2}}$
$\alpha_{\bar{1}}$	$\alpha_{\bar{1}}$	$\alpha_{\bar{2}}$	$\alpha_{\bar{0}}$
$\alpha_{\bar{2}}$	$\alpha_{\bar{2}}$	$\alpha_{\bar{0}}$	$\alpha_{\bar{1}}$

$\odot$	$\alpha_{\bar{0}}$	$\alpha_{\bar{1}}$	$\alpha_{\bar{2}}$
$\alpha_{\bar{0}}$	$\alpha_{\bar{0}}$	$\alpha_{\bar{0}}$	$\alpha_{\bar{0}}$
$\alpha_{\bar{1}}$	$\alpha_{\bar{0}}$	$\alpha_{\bar{1}}$	$\alpha_{\bar{2}}$
$\alpha_{\bar{2}}$	$\alpha_{\bar{0}}$	$\alpha_{\bar{2}}$	$\alpha_{\bar{1}}$

The isomorphism β (from Z_3 onto $\mathscr{R}$) corresponds $\bar{0}$ to $\alpha_{\bar{0}}$, $\bar{1}$ to $\alpha_{\bar{1}}$, and $\bar{2}$ to $\alpha_{\bar{2}}$. Notice that Z_3 and $\mathscr{R}$ differ only in the notation used to indicate the elements and operations.

EXERCISES

1. Verify that $\langle R, \oplus, \odot \rangle$ as given in 5.4.1 is a ring. (*Hint:* Each property follows immediately from the corresponding property in R; e.g., for commutativity of $\oplus$, we have $\alpha_a \oplus \alpha_b = \alpha_{a+b} = \alpha_{b+a} = \alpha_b \oplus \alpha_a$.)

†2. Recall that mappings are sets of ordered pairs and list these sets for the mappings $\alpha_{\bar{0}}$, $\alpha_{\bar{1}}$, and $\alpha_{\bar{2}}$ in the example of this section.

†3. In the proof of 5.4.1, where did we use the hypothesis that R has an identity e?

5.5   SUBRINGS AND IDEALS

The "ring in a ring" relationship, exemplified among the familiar number systems by Z and Q (also Q and F), leads us to the study of subrings just as the "group in a group" relationship led to the study of subgroups. Notice that between Z and Q we have

1. $Z \subseteq Q$,
2. the operation $+$ on Z is the operation $+$ on Q simply restricted to the elements of Z, and
3. the operation $\cdot$ on Z is the operation $\cdot$ on Q also restricted to the elements of Z.

This interesting relationship between the familiar rings initiates the second technique in the study of rings—namely, a study of some particular subsets of a ring. We simply require the subset to possess enough of the basic properties to be a ring itself.

5.5.1 DEFINITION. Let S be a nonempty subset of a ring R. S is a *subring* of R if S is a ring with respect to the operations on R restricted to S.

It would be a long procedure to have to verify each of the ring properties in order to show that a subset S is a subring. Fortunately, as with subgroups, a shortcut is provided by the following theorem.

5.5.2 THEOREM. Let S be a nonempty subset of a ring R. The following statements are equivalent.
1. S is a subring of R.
2. For all $a,b \in S$ we have $(a + b) \in S$, $-a \in S$, and $ab \in S$.
3. For all $a,b \in S$ we have $(a - b) \in S$ and $ab \in S$.

Proof: Clearly (1) implies (2). (2) implies (3) is Exercise 1 of this section. We show that (3) implies (1).

Restricting our attention to the additive structure, we have $\langle S, + \rangle$ is an abelian group by 3.4.3a in Section 4.5. Since $ab \in S$ for all $a,b \in S$, $\cdot$ restricted to S is an operation on S. Associativity for $\cdot$ and distributivity of $\cdot$ over $+$ are inherited from R. Thus S is a ring and hence a subring of R.▲

EXAMPLES

1. Let R be a ring. R itself is a subring of R, and $\{0\}$, together with $+$ and $\cdot$ as in R, is a subring. Both are referred to as *nonproper* subrings and $\{0\}$ is called the *trivial* subring of R.

2. Z is a subring of Q and Q is a subring of F. Is Z a subring of F?

3. $\{\overline{0},\overline{2},\overline{4}\}$ is a subring of Z_6.

4. $\{0,a\}$ is a subring of R in Example 7 of Section 5.1. Are $\{0,b\}$ and $\{0,c\}$ subrings of R?

5. Let $\eta: R_1 \to R_2$ be a ring homomorphism. $R_1\eta$ is a subring of R_2 (see 5.3.2).

6. In Section 5.1, the ring in Example 8 is a subring of the ring in Example 5.

These examples illustrate the fact that we must be careful not to assume much more than the basic ring properties in a subring without a detailed examination of the subring as a ring itself. For instance, an identity may or may not exist in a subring regardless of whether or not one exists in the ring. Consider R as in Example 7 of Section 5.1. R has no identity; yet the subring $\{0,a\}$ has an identity, namely, a. On the other hand, the ring of 2×2 matrices of real numbers has an identity, whereas the subring of 2×2 matrices of even integers (see Example 6) does not.

To complicate matters further, a subring of a ring with identity may have an identity that is different from the ring identity (an example will be given in Section 5.7). In addition, a subring of a noncommutative ring may or may not be commutative (see Example 4). It is easily shown, however, that every subring of a commutative ring is commutative.

A subring S of a ring is required to be closed with respect to multiplication (i.e., $\cdot$ restricted to S must be an operation on S). Some subrings barely satisfy this requirement, whereas others satisfy it above and beyond the bare necessities. Consider Z as a subring of Q. For $\frac{1}{2} \in Q$ and $3 \in Z$, the product $\frac{3}{2} \notin Z$. In this case, we must be sure that both a and b are in Z before we can conclude that $ab \in Z$ (although for particular choices of a and b, one can obtain a product that is in Z). However, if we consider the ring E of even integers as a subring of Z, then for any choice of one element in E and one element in Z, the product is in E. Here E is "superclosed," in a sense, with respect to the operation of multiplication. For two arbitrary elements in Z, it is sufficient to require that one of them be in E in order to guarantee that their product is in E.

This superclosure property sets subrings apart much like normality does with subgroups. The remainder of this section and the next section will discuss this analogy.

5.5.3 DEFINITION. Let I be a subring of a ring R. I is called an *ideal* of R if for each $a \in R$ and $b \in I$ we have $ab \in I$ and $ba \in I$.

The next result gives the usual test for determining whether or not a subset of a ring is an ideal. Like 5.5.2, it offers a shortcut when compared with the definition.

5.5.4 THEOREM. Let I be a nonempty subset of a ring R. I is an ideal if and only if
1. for all $a,b \in I$ we have $(a - b) \in I$, and
2. for all $a \in R$ and $b \in I$ we have $ab \in I$ and $ba \in I$.

Proof: Clearly if I is an ideal, then (1) and (2) hold. For the converse, suppose that conditions (1) and (2) hold for I. Then I is a subring by 5.5.2.3. Condition (2) is the superclosure property and is satisfied. Thus I is an ideal.▲

EXAMPLES

1. E, the ring of even integers, is an ideal of Z. Is E an ideal of Q?
2. $\{\overline{0},\overline{2},\overline{4}\}$ is an ideal of Z_6.
3. The nonproper subrings of a ring R (R and $\{0\}$) are ideals of R.

We remarked in Section 5.3 that a ring homomorphism, $\alpha: R \rightarrow R'$, is a group homomorphism from $\langle R, + \rangle$ to $\langle R', \oplus \rangle$. As such, α has a kernel. We now consider this same subset in the more sophisticated structure of a ring. The old definition (3.4.5) remains essentially unchanged. We are still looking at the set of pre-images of the "group identity" in R'.

5.5.5 DEFINITION. Let $\alpha: R \to R'$ be a ring homomorphism and let $0'$ be the additive identity in R'. The *kernel* of α is the set of all pre-images of $0'$ in R. Symbolically,

$$\ker \alpha = \{a \mid a \in R \text{ and } a\alpha = 0'\}.$$

In group theory we showed that the kernel of a group homomorphism is a normal subgroup. The corresponding result in ring theory is that the kernel of a ring homomorphism is an ideal.

5.5.6 THEOREM. Let α be a ring homomorphism. Then $\ker \alpha$ is an ideal of dom α.

Proof: Let $\alpha: R \to R'$ be a ring homomorphism and let $0'$ be the identity in R'. We appeal to 5.5.4. For $a,b \in \ker \alpha$ and $c \in R$, $(a - b)\alpha = (a + (-b))\alpha = a\alpha + (-b)\alpha = a\alpha + (-(b\alpha)) = a\alpha - b\alpha = 0' - 0' = 0'$ and $(ac)\alpha = (a\alpha)(c\alpha) = 0'(c\alpha) = 0'$. Similarly, $(ca)\alpha = 0'$. Therefore $(a - b) \in \ker \alpha$ and $ac, ca \in \ker \alpha$. Thus $\ker \alpha$ is an ideal of R.▲

5.5.7 THEOREM. Let α be an onto ring homomorphism. Then α is a ring isomorphism if and only if $\ker \alpha$ is the trivial subring of dom α.

Proof: Exercise 4 of this section.▲

EXERCISES

†1. Show that the only subring of Z that contains the identity is Z itself.

2. Verify that $\{0,a\}$ is a subring of R in Example 4. Is $\{0,a\}$ an ideal of R?

3. Let S be a subset of a ring R. Show that the following two statements of the superclosure property are equivalent.
   (a) For all $a,b \in R$, if $a \in S$, then $ab, ba \in S$.
   (b) For all $a,b \in R$, if $a \in S$ or $b \in S$, then $ab \in S$.

4. Prove 5.5.7.

5. Let a be a fixed element in a ring R. Show that $S = \{x \mid x \in R \text{ and } ax = 0\}$ is a subring of R but not necessarily an ideal of R. If R is commutative, is S an ideal of R?

6. Show that the intersection of a collection of subrings (ideals) of a ring R is a subring (ideal) of R.

†7. Construct an example to show that $S \cup T$ is not necessarily a subring when S and T are subrings.

8. Let R be a ring of order m (which means that R has m elements) and let S be a subring of R. Prove that the order of S divides m. (*Hint:* Recall Lagrange's theorem.)

9. Determine ker α in Example 1 of Section 5.3. Find ker β in Example 3 of Section 5.3.

†10. Show that every subring of Z is an ideal of Z.

11. Let $\alpha: R \to R'$ be a ring homomorphism and let S' be a subring of R'. Prove that S is a subring of R and ker $\alpha \subseteq S$, where $S = \{x \mid x \in R$ and $x\alpha \in S'\}$.

†12. Let I be an ideal of a ring R with identity e. Prove: if $e \in I$, then $I = R$.

13. Let A and B be ideals in ring R. Show that $A + B = \{a + b \mid a \in A$ and $b \in B\}$ is an ideal of R.

†14. Let a be a fixed element in a commutative ring R. Show that $aR = \{ar \mid r \in R\}$ is an ideal.

5.6   QUOTIENT RINGS AND A FUNDAMENTAL THEOREM

In group theory the concept of subgroup led to cosets, and consideration of normal subgroups led to the development of factor groups. The analogous sequence in ring theory proceeds from subrings to cosets and, then, considering subrings that are ideals, to quotient rings (the counterpart of factor groups). The definition of a coset remains unchanged except for the "dress up" needed because of the change of notation and the different structure. The additive notation for cosets may look a little strange, but remember that it is the additive structure in the ring which satisfies the group concept. Frequent reference to Section 4.5 should prove helpful in this section.

5.6.1 DEFINITION. Let S be a subring of a ring R. For each $x \in R$, $x + S = \{x + s \mid s \in S\}$ is called a *coset* of x and S in R.

There is no need to distinguish between left and right cosets in a ring because $\langle R, + \rangle$ is abelian and $x + S = S + x$ for all $x \in R$. $\langle S, + \rangle$ is a normal subgroup of $\langle R, + \rangle$. Therefore we have a corresponding factor group R/S and, by the Fundamental Theorem of Group Homomorphisms, a corresponding group homomorphism.

The group operation on R can be used on coset representatives to induce a group operation on R/S, the set of all cosets of S in R

$[(a + S) + (b + S) = (a + b) + S$; see 4.2.1a in 4.5]. By 4.2.4, we know that R/S, with this induced operation, is an abelian group. Now the question arises, can we use the operation of multiplication on R in the same natural way $[(a + S) \cdot (b + S) = ab + S]$ to obtain an operation of multiplication on R/S that will possess the required ring properties? Let us examine R/S with $+$ and $\cdot$ defined in this natural way for two specific examples.

EXAMPLES

1. Consider the subring E of even integers of the ring Z. There are only two cosets; $0 + E = E$ and $1 + E$, the set of odd integers. The tables for $+$ and $\cdot$ on Z/E are

$+$	E	$1 + E$
E	E	$1 + E$
$1 + E$	$1 + E$	E

$\cdot$	E	$1 + E$
E	E	E
$1 + E$	E	$1 + E$

By tedious checking we can verify that $\langle Z/E, +, \cdot \rangle$ is a ring (or by showing that $Z/E \approx Z_2$).

2. Consider Z as a subring of Q. $(\frac{1}{2} + Z)(\frac{1}{3} + Z)$ is defined to be $(\frac{1}{6} + Z)$. But $\frac{1}{2} + Z = \frac{3}{2} + Z$ and $\frac{1}{3} + Z = \frac{7}{3} + Z$ (see 3.5.3a in 4.5). Therefore
$$(\frac{1}{2} + Z)(\frac{1}{3} + Z) = (\frac{3}{2} + Z)(\frac{7}{3} + Z) = \frac{7}{2} + Z = \frac{1}{2} + Z$$
and $\cdot$ is not a binary operation on Q/Z (since $\frac{1}{6} + Z \neq \frac{1}{2} + Z$). Conclusion: Q/Z cannot be "made into a ring" in this manner even though $\langle Q/Z, + \rangle$ is an abelian group.

The difference between these two examples, as you have probably already noticed, is that the subring in Example 1 is an ideal, whereas in Example 2 the subring is not an ideal. Since the ideal is the ring substructure that corresponds to normal subgroups, we might expect R/S to be a ring only when S is an ideal. Further motivation for restricting our attention to ideals, when considering R/S, is that the Fundamental Theorem of Group Homomorphisms associates factor groups with kernels of homomorphisms and we have already established that the kernel of a ring homomorphism is an ideal.

5.6.2 THEOREM. If I is an ideal in the ring R and R/I is the collection of all cosets of I in R, then $\langle R/I, +, \cdot \rangle$ is a ring when $+$ and $\cdot$ are defined by

$$(a + I) + (b + I) = (a + b) + I,$$

and

$$(a + I) \cdot (b + I) = ab + I.$$

R/I is called a *quotient ring*.

Proof: By 4.2.4, $\langle R/I, + \rangle$ is an abelian group. To prove that the induced multiplication is an operation on R/I, it suffices to show that if $x \in (a + I)$ and $y \in (b + I)$, then $xy \in (ab + I)$. Suppose that $x \in (a + I)$ and $y \in (b + I)$. Then there exist $k_1, k_2 \in I$ such that $x = a + k_1$ and $y = b + k_2$. Now $xy = (a + k_1)(b + k_2) = ab + ak_2 + bk_1 + k_1k_2$. I is an ideal. Thus $(ak_2 + bk_1 + k_1k_2) \in I$ and $xy \in (ab + I)$.

Associativity and distributivity follow immediately from the corresponding properties in R. Hence R/I is a ring. ▲

EXAMPLE. Consider Z_6. $I = \{\bar{0}, \bar{3}\}$ is an ideal of Z_6. The elements of the quotient ring Z_6/I are $\bar{0} + I$ (or just I), $\bar{1} + I$, and $\bar{2} + I$ (note that $\bar{3} + I = I$, $\bar{4} + I = \bar{1} + I$, and $\bar{5} + I = \bar{2} + I$). Construct the addition and multiplication tables for Z_6/I.

Associated with each normal subgroup N of a group G is the natural map v from G to G/N defined by $gv = g + N$. This map is a group homomorphism and ker $v = N$ (see 4.2.8). This natural map, when considered relative to a ring R with an ideal I, turns out to be a ring homomorphism whose kernel is I.

The next two results, along with 5.5.6, establish the correspondence between ideals of a ring R and homomorphic images of R (up to isomorphism).

5.6.3 THEOREM. Let I be an ideal of a ring R. If $v: R \to R/I$ is the natural map defined by $rv = r + I$ for all $r \in R$, then v is an onto ring homomorphism and ker $v = I$.

Proof: Exercise 1 of this section. ▲

5.6.4 THEOREM. If $\alpha: R \to R'$ is an onto ring homomorphism and ker $\alpha = I$, then $R/I \approx R'$.

Proof: By 4.3.1, $\langle R/I, + \rangle \approx \langle R', + \rangle$ via the map β defined by $(r + I)\beta = r\alpha$ for all $(r + I) \in R/I$ (the proof of 4.3.1 illustrates why we defined β in this way). Since β is 1–1, onto, and preserves addition, we need only show that β preserves multiplication. Let $x + I$, $y + I$ $\in R/I$. Then

$$
\begin{aligned}
[(x + I)(y + I)]\beta &= (xy + I)\beta \\
&= (xy)\alpha \\
&= (x\alpha)(y\alpha) \\
&= (x + I)\beta(y + I)\beta. \blacktriangle
\end{aligned}
$$

As in Section 4.3, the results given by 5.5.6, 5.6.3, and 5.6.4 are each regarded as part of a fundamental theorem that shows the relationship between ideals, ring homomorphisms, and homomorphic images of rings.

5.6.5 THE FUNDAMENTAL THEOREM OF RING HOMO-MORPHISMS. Associated with each ring homomorphism of a ring R is an ideal of R, and associated with each ideal I of R is a ring homomorphism of R with I as its kernel. Furthermore, the only homomorphic images of R are the quotient rings (up to isomorphism).

EXERCISES

1. Prove 5.6.3.

†2. Let R be a ring. Exhibit all homomorphic images of R (up to isomorphism) if the number of elements in R is prime (see Exercise 6 in 4.3).

3. Let R and R' be rings. Verify that the cartesian product $R \times R'$ with $\oplus$ and $\odot$ defined by $(x, x') \oplus (y, y') = (x + y, x' + y')$ and $(x, x') \odot (y, y') = (xy, x'y')$ is a ring. This ring is called the *direct sum* of R and R' and is denoted by $R \oplus R'$.

4. (From 3) Show that R is isomorphic to an ideal of $R \oplus R'$. (*Hint:* Consider $\{(r, 0' | r \in R\}$.)

5. If I is an ideal of a ring R, prove that there is a 1–1 correspondence between the subrings of R/I and the subrings of R that contain I. (*Hint:* See 4.2.5.)

6. If I is an ideal of a ring R, prove that there is a 1–1 correspondence between the ideals of R/I and the ideals of R that contain I. (*Hint:* See 4.2.6.)

7. If I is an ideal of a commutative ring R, show that R/I is commutative.

†8. If I is an ideal of a ring R with identity, show that R/I is a ring with identity.

5.7   IMBEDDING A RING IN A RING WITH IDENTITY

Several examples of rings that we have examined do not have an identity. In this section we will show that each of these rings can be considered (via an isomorphism) as a subring of a ring that does have an identity. Informally, a ring R can be imbedded in a ring R' if R can be considered as a subring of R'. The concept of imbedding is used with algebraic systems in general as well as with rings.

5.7.1 DEFINITION. An algebraic system S is said to be *imbedded* in an algebraic system S' if S is isomorphic to a subsystem of S'. Furthermore, if S can be imbedded in S', we say that S' is an *extension* of S.

EXAMPLES

1. Exercises 7 and 8 in 4.3 establish that the direct product $G \times G'$ of groups G and G' is a group and that G is isomorphic to a normal subgroup of $G \times G'$. Therefore G is imbedded in $G \times G'$ and $G \times G'$ is an extension of G.

2. Exercise 9 of Section 4.3 provides another example. When H and K are normal subgroups of G, then $G/(H \cap K)$ is isomorphic to a subgroup of $G/H \times G/K$; that is, $G/(H \cap K)$ is imbedded in $G/H \times G/K$.

3. Exercises 3 and 4 in Section 5.6 establish that the direct sum ring $R \overset{*}{\oplus} R'$ is an extension of the ring R.

In Section 3.7 we noted that in an additive group the definition for "powers of elements" becomes a definition for "multiples of elements." Therefore in a ring we have:

1. the zero multiple of a ring element r is 0, the additive identity for R,

2. a positive multiple m of r signifies repeated addition of r m times, and

3. a negative multiple $-m$ of r corresponds to a positive multiple m of the element $(-r)$.

It is important to keep the notation meaningful. For example, when we see the zero multiple of r—that is, $0r = 0$—remember that the 0 on the left is the integer zero and the 0 on the right is the additive identity for R. Also, if $m \in Z$, and $r \in R$, then mr is not multiplication in R but repeated addition of the ring element r. The integer m cannot be considered as a ring element (except in the special case when $Z \subseteq R$ or $R \subseteq Z$). This concept of integral multiples of a ring element is used in defining the operation of multiplication in our extension ring in the following.

5.7.2 THEOREM. Any ring R can be imbedded in a ring with identity.

Proof: We first construct the extension ring and then display the subring that is a "copy" of R.

The construction. Consider the cartesian product set $R \times Z$. For (x,m), $(y,n) \in R \times Z$, define $\oplus$ and $\odot$ as follows:
$$(x,m) \oplus (y,n) = (x + y, m + n)$$
and
$$(x,m) \odot (y,n) = (xy + nx + my, mn).$$
Exercise 1 verifies that $\langle R \times Z, \oplus, \odot \rangle$ is a ring with identity. The identity is $(0,1)$.

The subring. The natural subset of $R \times Z$ to consider is $R' = \{(r,0) \mid r \in R\}$. Exercise 2 shows that R' is a subring of $R \times Z$.

The isomorphism. The obvious mapping to consider is $\beta: R \to R'$ defined by $r\beta = (r,0)$ for all $r \in R$. Clearly β is 1-1 and onto. To show that β preserves both operations, let $x,y \in R$. Then
$$(x + y)\beta = (x + y, 0) = (x,0) \oplus (y,0) = x\beta \oplus y\beta$$
and
$$(xy)\beta = (xy,0) = (x,0) \odot (y,0) = x\beta \odot y\beta. \blacktriangle$$

We are now in a position to construct easily a subring of a ring with identity where the subring has an identity different from the ring identity. Recall that in Section 5.5 we deferred such an example to this section. Let R in the proof of 5.7.2 be a ring with identity e. Then $(e,0)$ is the identity for R'. Since $(0,1)$ is the identity for $R \times Z$, R' is a subring of $R \times Z$ that has an identity different from the identity of $R \times Z$.

In Section 5.4 we showed that each ring with identity is isomorphic to a ring of mappings (in particular, a ring of group homomorphisms). With the aid of 5.7.2 we can obtain the more general result.

5.7.3 THEOREM. Every ring R is isomorphic to a ring of mappings, in particular, to a ring of group homomorphisms from an abelian group to itself.

Proof: Let R' be the copy of R in the extension ring $R \times Z$ (see the proof of 5.7.2). By 5.4.1, since $R \times Z$ has an identity, $R \times Z \approx \mathscr{R}$ a ring of group homomorphisms from an abelian group to itself. Let $\alpha: R \times Z \to \mathscr{R}$ be an isomorphism and $\eta: R' \to \mathscr{R}$ be the restriction of α to R; that is, $(r,0)\eta = (r,0)\alpha$ for all $(r,0) \in R'$. Recall that $\beta: R \to R'$ as defined in 5.7.2 is an isomorphism. This information is shown in the diagram. $R'\eta$ is a subring of $\mathscr{R}$ and $\eta: R' \to R'\eta$ is an isomorphism.

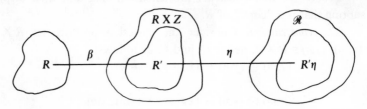

Therefore $\beta\eta: R \to R'\eta$ is an isomorphism and $R \approx R'\eta$. ▲

EXERCISES

Exercises 1 through 5 refer to the proof of 5.7.2.

1. Verify that $R \times Z$ is a commutative ring with identity.

2. Show that R' is a subring of $R \times Z$.

†3. If R has an identity e, show that there exists an element $(x,n) \in R \times Z$ with $(x,n) \neq (0,0)$ such that $(e,0) \odot (x,n) = (0,0)$.

4. Show that E, the ring of even integers, can be imbedded in a ring with identity in at least two different ways. (*Hint:* $E \approx E' \subseteq E \times Z$ and $E \subseteq Z$.)

5. Show that R' is an ideal. This establishes that any ring R can be imbedded as an ideal in a ring with identity.

6. Let R be a ring. If $\langle R, + \rangle$ is cyclic, prove that R is a commutative ring.

An element a in a ring R is *nilpotent* if there exists a positive integer n such that $a^n = 0$. Exercises 7 through 10 involve nilpotent elements.

7. If x is nilpotent in a ring R, $x \neq 0$, and R has more than two elements, prove that there exists $y \in R$ such that y is nilpotent, $y \neq 0$, and $y \neq x$.

†8. If a is a nilpotent element in a commutative ring R, show that ab is nilpotent for each $b \in R$.

9. If $b^2 = 0$ in a commutative ring with identity and a^{-1} exists. (multiplicative inverse of a), show that $(a + b)^{-1}$ exists. Does $(a + b)^{-1}$ exist if $b^3 = 0$? If b is nilpotent?

10. Show that in a commutative ring the set of all nilpotent elements is an ideal.

5.8   PRINCIPAL IDEALS

Ideal theory is an extremely important part of ring theory. The existence or nonexistence of certain of these special subsets yields much information about the structure of a ring. Their assistance in determining the structure of some particular quotient rings will be considered in the next chapter. Ideals are valuable subsets of all rings, but for our purposes in this section (in fact, throughout the remainder of the text) we will consider only commutative rings.

The first type of ideal we consider is the "smallest" ideal containing a given subset of a ring. For any subset A, of a ring R, we can find ideals of R that contain A. R itself is one such ideal. Let $\mathscr{I} = \{I \mid A \subseteq I \text{ and } I \text{ is an ideal of } R\}$. By Exercise 6 in Section 5.5, $\cap \mathscr{I}$ is an ideal of R and clearly $\cap \mathscr{I}$ is the smallest ideal in R that contains A—smallest in the sense that if S is any ideal of R that contains A, then $\cap \mathscr{I} \subseteq S$.

5.8.1 DEFINITION. Let R be a ring, $A \subseteq R$, and $\mathscr{I} = \{I \mid A \subseteq I \text{ and } I \text{ is an ideal of } R\}$. Then $\cap \mathscr{I}$ is called the ideal *generated* by A and is denoted by (A) or $(a_1, a_2, \ldots, a_n)$ when $A = \{a_1, a_2, \ldots, a_n\}$ is finite. Furthermore, if $A = \{a\}$, we call (a) the *principal ideal generated by a in R*.

Keeping in mind the fact that (a) is a subring of R with a few extra properties, it is a worthy exercise in notation to determine the form of each element in (a).

5.8.2 THEOREM. Let a be an element of a commutative ring R. Then each element of (a) is of the form $na + ra$, where $n \in Z$ and $r \in R$. Moreover, if R has an identity, then (a) consists of all elements of the form ra for $r \in R$.

Proof: Let $I_a = \{na + ra \mid n \in Z \text{ and } r \in R\}$. We wish to show that $(a) = I_a$. First, we will verify that I_a is an ideal. Since

$a = 1a + 0a$ ($1 \in Z$ and $0 \in R$), $a \in I_a$. Now let $x, y \in I_a$ and $r \in R$. For $x = n_1 a + r_1 a$ and $y = n_2 a + r_2 a$, we have $x - y = (n_1 - n_2)a + (r_1 - r_2)a$. Since $(n_1 - n_2) \in Z$ and $(r_1 - r_2) \in R$, $(x - y) \in I_a$. For superclosure,

$$xr = rx = r(n_1 a + r_1 a)$$
$$= (n_1 r)a + (rr_1)a$$
$$= (n_1 r + rr_1)a$$
$$= 0 \cdot a + (n_1 r + rr_1)a.$$

Since $(n_1 r + rr_1) \in R$, we have xr in the desired form. Thus $xr \in I_a$, and I_a is an ideal containing a, so $(a) \subseteq I_a$ since (a) is the intersection of all ideals containing a.

We now show that $I_a \subseteq (a)$. (a), as a subring, is a group under $+$ and hence must contain all multiples of a; that is, $na \in (a)$ for all $n \in Z$. As an ideal, superclosure guarantees that all elements of the form ra for $r \in R$ are in (a) since $a \in (a)$. Addition is an operation on (a); therefore all elements of the form $na + ra$ must belong to (a) and $I_a \subseteq (a)$.

If R contains an identity e, then elements of the form $na + ra$ can be written as follows:

$$na + ra = n(ea) + ra$$
$$= (ne)a + ra$$
$$= (ne + r)a$$
$$= r'a,$$

where $r' = ne + r$. Therefore in a commutative ring with identity, the principal ideal (a) consists precisely of the ring multiples of a; that is, $(a) = \{ra \mid r \in R\}$. ▲

EXAMPLES

1. In Z_6, $(\bar{1}) = Z_6$, $(\bar{2}) = \{\bar{0}, \bar{2}, \bar{4}\}$, $(\bar{3}) = \{\bar{0}, \bar{3}\}$, and $(\bar{2}, \bar{3}) = Z_6$.
2. In Z, $(1) = Z$ and $(4) = \{n \cdot 4 \mid n \in Z\}$. In Q,
$$(4) = \{q \cdot 4 \mid q \in Q\} = Q.$$

However, in the ring of even integers E,
$$(4) \neq \{m \cdot 4 \mid m \in E\}$$

since E has no identity. To see this, note that $12 \in (4)$ and
$$12 \notin \{m \cdot 4 \mid m \in E\}.$$

Why is $12 \in (4)$?

Let a be an element of a commutative ring R. The ideal (a) contains all positive powers a^m of a, since for $m = 1$, $a^1 = 1a + 0a$

and for all integers $m > 1$, $a^m = a^{m-1} \cdot a = 0a + ra$, where $0 \in Z$ and $r = a^{m-1} \in R$. If R has an identity e, then (a) may or may not contain $a^0 = e$. In the ring of integers (2) does not contain the identity, whereas $(1) = Z$ does. Furthermore, (a) may fail to contain negative powers of a even when they exist. Recall that for a positive integer n, a^{-n} denotes the multiplicative inverse of a^n.

Ideals other than principal ideals exist (see Exercise 5 of this section); however, in Z every ideal is a principal ideal. Rings in which every ideal is principal are called *principal ideal rings* (PIR).

5.8.3 THEOREM. The ring of integers is a PIR.

Proof: Let I be an ideal of Z. If I is the zero ideal, then $I = (0)$. Suppose that $I \neq (0)$. $I \subseteq Z$; therefore I has a smallest positive integer, say a. By 5.8.2, $(a) = \{na \,|\, n \in Z\}$. We proceed to show that $I = (a)$. Clearly, since $a \in I$, we have $(a) \subseteq I$. Let $x \in I$. By the division algorithm (2.7.1), there exist integers q and r, with $0 \leq r < a$, such that $x = qa + r$. Now $r = x - qa$, $x \in I$, and $qa \in I$; thus $r \in I$. Since $0 \leq r < a$ and a is the smallest positive integer in I, we must have $r = 0$. Hence $x = qa \in (a)$ and $I \subseteq (a)$.▲

EXERCISES

1. Let a and b be elements of a ring with identity. Give a counterexample to disprove the following: if $(a) = (b)$, then $a = b$.

†2. Let R be a commutative ring with identity, I an ideal of R, and $a \in I$. Prove that if a has a multiplicative inverse, then $I = R$.

3. Find positive integers m and n so that $(m) \cup (n)$ is not an ideal of Z.

†4. Find a unique positive generator for each of the following ideals of Z:
   (a) $(1,2)$; (b) $(2,3)$; (c) $(2,4)$; (d) $(5,10)$; (e) $(2,5,10)$.

5. Consider $\langle P(Z), +, \cdot \rangle$ (see Example 4 in 5.1). Let $I = \{A \,|\, A \in P(Z)$ and A is finite$\}$. Prove:
   (a) I is an ideal of $P(Z)$,
   (b) I is not a principal ideal.

6. Let R be a commutative ring with identity and $a, b \in R$. Show that:
   (a) $\{a,b\} \subseteq (a) + (b)$, where $(a) + (b) = \{x + y \,|\, x \in (a)$ and $y \in (b)\}$, and
   (b) $(a) + (b) = (a,b)$.

7. Consider the direct sum ring $Z \oplus Z$ (see Exercise 3 in 5.6). Let A and B denote the ideals generated by $(1,0)$ and $(0,2)$ respectively. Is $A + B$ a principal ideal where $A + B = \{a + b \mid a \in A \text{ and } b \in B\}$? Verify your answer.

†8. Give an alternate proof of 5.8.3, using the fact that each ideal of Z is an additive subgroup that is cyclic.

9. Show that Z_m is a PIR for each integer $m \geq 1$.

CHAPTER 6

INTEGRAL DOMAINS
AND FIELDS

Most of the definitions and theorems that we have considered arise from attempts to generalize certain useful properties of the familiar number systems. We started with the group concept and, then, by requiring additional familiar properties, developed the structure of a ring. This chapter examines algebraic systems that possess, in addition to the ring properties, one or more properties analogous to the following:

1. If $a,b \in Z$ with $ab = 0$, then $a = 0$ or $b = 0$,
2. each nonzero rational number has a multiplicative inverse, and
3. the system of integers possesses an order relation.

6.1   DEFINITION OF AN INTEGRAL DOMAIN AND A FIELD

The first additional property that we choose to focus our attention on is related to zero products. Exercise 7(b) in Section 5.1 shows that the familiar property "$ab = 0$ implies $a = 0$ or $b = 0$" does not have to hold in the general ring. Nonzero ring elements that violate this property motivate the need for the term zero divisor.

6.1.1 DEFINITION. A nonzero element a in a commutative ring R is called a *zero divisor* if there exists a nonzero element $b \in R$ such that $ab = 0$.

Certainly if a is a zero divisor, then so is b. A more detailed study distinguishes between left and right zero divisors and, of course, does not require that the ring be commutative. We require commutativity and therefore have no way of distinguishing between left and right zero divisors.

Z, Q, and F do not contain any zero divisors; however, Z_6 has three. Can you list them? For ring elements that are not divisors of zero, the operation of multiplication possesses an important and familiar property.

6.1.2 THE CANCELLATION LAW FOR MULTIPLICATION. Let a, b, and c be elements of a commutative ring R. If $a \neq 0$ is not a zero divisor and $ab = ac$, then $b = c$.

Proof: If $ab = ac$, then $ab - ac = 0$ and $a(b - c) = 0$. Now $a \neq 0$, and a is not a zero divisor; therefore $b - c = 0$ and $b = c$. ▲

6.1.3 DEFINITION. An *integral domain* is a commutative ring with identity that has no zero divisors.

The property "$ab = 0$ implies $a = 0$ or $b = 0$" is required by definition to hold for all elements a and b in an integral domain. The most familiar example of an integral domain is of course Z. Q and F are also integral domains. However, the ring of even integers and Z_6 are not integral domains. Why?

It is immediate from 6.1.2 that the cancellation law for multiplication holds for all nonzero elements in an integral domain. The converse is also true.

6.1.4 THEOREM. Let R be a commutative ring with identity. If the cancellation law for multiplication holds for all nonzero elements of R, then R is an integral domain.

Proof: Let $a \in R$ and $a \neq 0$. Suppose that $b \in R$ is such that $ab = 0$. Then $ab = a0$ since $a0 = 0$. By the cancellation law for multiplication, since $a \neq 0$, we have $b = 0$. Hence a is not a zero divisor, and since a is an arbitrary nonzero element, R is an integral domain. ▲

The rational and real number systems have a structure only slightly more sophisticated than an integral domain. The difference is that every nonzero element has a multiplicative inverse. Systems having this additional property are called fields. Formally, we have the following definition.

6.1.5 DEFINITION. A *field* is a commutative ring with identity in which the nonzero elements form a multiplicative group.

Since the nonzero elements must form a group, a field must contain at least two elements and the identity e must be different from 0. Why? Q and F are the most familiar examples of a field. Another example is Z_2 (in fact, Z_p is a field, for each prime p; this point will be shown in 6.3). Each example of a field that we have discussed is also an integral domain. This is always the case.

6.1.6 LEMMA. Let R be a commutative ring with identity. If $a \in R$ has an inverse (multiplicative), then a is not a zero divisor.

Proof: Exercise 1 of this section. ▲

In a field, every nonzero element must have an inverse. Therefore no element can be a zero divisor, and we have

6.1.7 THEOREM. Every field is an integral domain.

There are many examples of integral domains that are not fields (the ring $\langle Z, +, \cdot \rangle$ is one such example). Therefore the converse of 6.1.7 cannot hold. We can, however, obtain a partial converse.

6.1.8 THEOREM. A finite integral domain with at least two elements is a field.

Proof: Exercise 2 of this section. ▲

EXERCISES

†1.   Prove 6.1.6.

2.   Prove 6.1.8. (*Hint:* See the proof of 3.4.4.)

3.   Let D be an integral domain. If A is a subring of D containing the identity, prove that A is an integral domain.

4.   Why can 0 not have an inverse in a field?

†5.   Let T be a field. Show that for each $a, b \in T$ with $a \neq 0$, there is a unique $x \in T$ such that $ax = b$.

6.   Let $T = F \times F$. Define $\oplus$ and $\odot$ on T as follows: for all $(a,b), (c,d) \in T$,
$$(a,b) \oplus (c,d) = (a + c, b + d),$$
and
$$(a,b) \odot (c,d) = (ac - bd, ad + bc).$$
Show that $\langle T, \oplus, \odot \rangle$ is a field.

7.   Show that a field contains no proper ideals. (*Hint:* Use the result of Exercise 2 in 5.8.)

8.   Let $\eta: T \to T'$ be a homomorphism from the field T onto the field T'. Show that η is either trivial (maps everything to $0' \in T'$) or an isomorphism. (*Hint:* Use Exercise 7 and then the Fundamental Theorem of Ring Homomorphisms.)

†9.   Show that in a field every ideal is a principal ideal.

10.   Prove that the only nilpotent element (see Exercises 7 through 10 in 5.7) in an integral domain is the additive identity 0.

11.   An element a of a ring is *idempotent* if $a^2 = a$. Prove that if a is idempotent in an integral domain D and $a \neq 0$, then a is the identity e of D.

†12.   Let $\mathscr{D}$, $\mathscr{T}$, and $\mathscr{F}$ denote the set of all integral domains, the set of all fields, and the set of all finite integral domains respectively. Show, via a venn diagram, the containment relationship between $\mathscr{D}, \mathscr{T}$, and $\mathscr{F}$.

†13.   If a is a zero divisor in a commutative ring R and $b \in R$ so that $ab \neq 0$, prove that ab is a zero divisor.

6.2   SPECIAL ELEMENTS

In the integers we say that 3 divides 21 because there is an integer 7 so that $3 \cdot 7 = 21$. By similar reasoning, 2 does not divide 21, since there is no integer k such that $2k = 21$. The concept of divisor or factor is directly related to the operation of multiplication and to the existence of "complementary factors" (3 and 7 complement each other as factors of 21). We generalize this concept to an arbitrary integral domain.

6.2.1 DEFINITION. Let D be an integral domain. For $a, b \in D$, *a divides b* (or *a is a factor of b*) if there exists $c \in D$ such that $ac = b$. Symbolically, we write $a \mid b$ if a divides b and $a \nmid b$ if a does not divide b.

The concept of a divisor (factor) of an element in an integral domain is defined so that we have a direct analogy to the same concept in Z. Divisors exist in more sophisticated structures like Q and F. However, their study in such structures is not very fruitful, since in a field any two nonzero elements divide one another. For this reason, we will restrict our attention to integral domains when studying divisors.

The next theorem lists some basic results that follow immediately from 6.2.1. They are deemed important enough to list in a theorem because we actually chose the definition to ensure that they hold.

6.2.2 THEOREM. Let D be an integral domain with identity e. Then for all $a \in D$,

1. $\pm a \mid a$ (i.e., $a \mid a$ and $-a \mid a$),
2. $\pm a \mid 0$,
3. $\pm e \mid a$.

Proof: Exercise 1 of this section.▲

It occasionally happens that a collection of elements in an integral domain will divide each other. In the integers, for instance, 5 and -5 divide each other. Collections of elements with this property are helpful in obtaining information about integral domains.

6.2.3 DEFINITION. Let D be an integral domain with identity *e.* For $a,b \in D$, a is an *associate* of b if $a \mid b$ and $b \mid a$. An associate of the identity *e* is called a *unit*.

Certainly if a is an associate of b, then b is an associate of a and we may refer to a and b as associates. Thus a and e are associates when a is a unit. In Z, 5 is an associate of -5 and the only units are 1 and -1.

The elementary properties possessed by associates and units are given in the collection of results that follows.

6.2.4 THEOREM. Let D be an integral domain. The following statements are equivalent.
1. u is a unit in D.
2. u has an inverse in D.
3. $u \mid a$ for each $a \in R$.
4. $(u) = D$.

Proof: Exercise 2 of this section. ▲

6.2.5 THEOREM. Let D be an integral domain. For $a,b \in D$, a and b are associates if and only if there exists a unit $u \in D$ so that $au = b$.

Proof: Exercise 6 of this section. ▲

The associate relationship between elements induces the equality relation between their corresponding principal ideals. This is one of the ways associates help in investigating the structure of an integral domain.

6.2.6 THEOREM. In an integral domain D, a and b are associates if and only if $(a) = (b)$.

Proof: Suppose that a and b are associates. Recall that $(b) = \{xb \mid x \in D\}$ and $(a) = \{ya \mid y \in D\}$. Let $s \in (b)$. Then $s = d_1 b$ for some $d_1 \in D$. Now $a \mid b$; thus there exists $d_2 \in D$ such that $d_2 a = b$. $s = d_1 b = d_1(d_2 a) = (d_1 d_2)a \in (a)$. Therefore $(b) \subseteq (a)$. Similarly, using the fact that $b \mid a$, $(a) \subseteq (b)$. Hence $(a) = (b)$.

The converse is obvious from the form of elements in (a) and (b). ▲

We pursue the concept of divisor further by introducing the term common divisor.

6.2.7 DEFINITION. Let a, b, and d be elements in an integral domain. Then d is a *common divisor* of a and b if $d\,|\,a$ and $d\,|\,b$. Furthermore, if every common divisor of a and b is also a divisor of d, then d is a *greatest common divisor* (GCD) of a and b.

The common divisors in Z of 24 and 36 are ± 2, ± 3, ± 4, ± 6, and ± 12. Since all common divisors divide ± 12, both 12 and -12 are GCDs of 24 and 36. Notice that the word "greatest" in GCD is used without reference to the relation $<$ on Z. We should expect this, since integral domains, in general, do not have a relation analogous to $<$ on Z. This fact will be established in Section 6.5.

In an integral domain with only principal ideals, we can express each GCD of a and b as a linear combination of a and b. An integral domain in which every ideal is principal is called a *principal ideal domain* (PID). The domain of integers, of course, is an example of a PID.

6.2.8 THEOREM. Let D be a PID and $a,b \in D$. Then there exist $x,y,d \in D$ such that $d = ax + by$ and d is a GCD of a and b.

Proof: Consider (a,b), the ideal generated by a and b. D is a PID; therefore there is some $d \in D$ with $(d) = (a,b)$. Now $d \in (a,b)$; thus there exist $x,y \in D$ such that $d = ax + by$. Since $a,b \in (d)$, $a = sd$ and $b = td$ for suitable $s,t \in D$. Hence d is a common divisor of a and b. To show that d is a GCD of a and b, let c be any common divisor of a and b. Then $a = cg$ and $b = ch$ for some $g,h \in D$. $d = ax + by$ $= (cg)x + (ch)y = c(gx + hy)$, so $c\,|\,d$ and d is a GCD of a and b. ▲

This theorem also tells us that every pair of elements in a PID has a GCD. More than one GCD usually exists for a given pair of elements. For instance, 6 and -6 are different GCDs of 12 and 18 in Z (also see Exercise 10).

6.2.9 COROLLARY. Let D be a PID. If d is any GCD of a and b, then there exist $x,y \in D$, so that $d = ax + by$.

Proof: Exercise 7 of this section. ▲

It is natural to follow the divisor concept with that of a prime element. Prime integers are those integers p, different from 0 and ± 1, whose only divisors are ± 1 and $\pm p$. Noting that ± 1 are the only units in Z and that $\pm p$ are the only associates of p in Z, we generalize the concept of prime to an arbitrary domain.

6.2.10 **DEFINITION.** An element p in an integral domain D is a *prime element* if $p \neq 0$, p is not a unit, and the only divisors of p are the units and associates of p. If p is not zero, not a unit, and not a prime, then p is called a *composite* in D.

A nonzero element in an integral domain is either a unit, a prime, or a composite. $\{\pm 1\}$, $\{\pm 2, \pm 3, \pm 5, \pm 7, \pm 11, \pm 13, \pm 17, \pm 19, \pm 23, \ldots\}$, and $\{\pm 4, \pm 6, \pm 8, \pm 9, \pm 10, \ldots\}$ are, respectively, the set of units, the set of primes, and the set of composites of Z.

We now prove a familiar property of prime integers in the setting of a PID.

6.2.11 **THEOREM.** Let D be a PID and $a,b \in D$. If p is a prime in D and $p \mid ab$, then $p \mid a$ or $p \mid b$.

Proof: Consider (p,a). Since D is a PID, we have $(p,a) = (d)$ for some $d \in D$. Now $p \in (d)$. Thus $p = hd$ for some $h \in D$. p is prime, so its only divisors are units and associates of p. We now consider two cases.

Case 1. Suppose that d is an associate of p. Then $(p,a) = (d) = (p)$ by 6.2.6. Now $a \in (p)$; therefore $p \mid a$.

Case 2. Suppose that d is a unit. Then, by 6.2.4, d^{-1} exists and we have $e = d^{-1}d \in (d) = (p,a)$—that is, $e \in (p,a)$. So there exist $x,y \in D$ such that $e = px + ay$. Finally, $p \mid ab$; therefore $ab = pk$ for some $k \in D$ and $b = eb = (px + ay)b = pxb + aby = pxb + pky = p(xb + ky)$. Hence $p \mid b$. ▲

EXERCISES

1. Prove 6.2.2.
†2. Prove 6.2.4.
3. Show that the set U of all units in an integral domain is a group with respect to multiplication.

4. Let a, b, and c belong to an integral domain D. If $a \mid b$ and $b \mid c$, prove the following: (a) $a \mid c$, (b) $a \mid (b + c)$, and (c) $a \mid bx$ for all $x \in D$.

5. Show that the relation "is an associate of" in an integral domain is an equivalence relation.

†6. Prove 6.2.5.

7. Prove 6.2.9.

8. Show that any two nonzero elements in a field are associates.

9. In 6.2.2, notice that we allow 0 to divide 0. Does 0 divide any other element of D?

10. Let D be a PID and $a,b,c,d \in D$ so that d is a GCD of a and b. Prove that c is a GCD of a and b if and only if c and d are associates.

†11. Let D be the ring of gaussian integers (see Exercise 11 in 5.1). Show that D is an integral domain. Given that the only units in D are $(1,0)$, $(-1,0)$, $(0,1)$, and $(0,-1)$, find all the associates of $(3,4)$.

†12. Show that neither prime nor composite elements exist in a field.

†13. What does Exercise 12 say about the "primeness" of 3?

14. Let D be an integral domain and $a,b \in D$. Prove that $a \mid b$ if and only if $(b) \subseteq (a)$.

15. Let $D = \{a + b\sqrt{2} \mid a,b \in Z\}$. Using ordinary addition and multiplication show that:

    (a) D is an integral domain, and

    (b) 2 is not a prime in D.

†16. Let D be a PID and $a,b,d \in D$. Show that d is a GCD for a and b if and only if $(d) = (a,b)$.

17. Let D be an integral domain and let a and b be nonzero elements of D. Prove:

    (a) $(ab) \subseteq (a)$,

    (b) $(ab) \subset (a)$ if a and b are not units.

6.3 MORE ON IDEALS AND QUOTIENT RINGS

In our initial study of ideals, we considered them basically as particular subsets of a ring. Attention is now directed to the types of ideals that determine the structure of their corresponding quotient rings.

The first type of ideal we consider occurs as a generalization of a principal ideal of Z with a prime generator. Recall that for all $a \in Z$, (a) has the superclosure property—that is, if $x \in (a)$ and $y \in Z$, then $xy \in (a)$. If a is a prime, we can obtain the converse of superclosure.

6.3.1 LEMMA. Let $a,b,p \in Z$ and let p be a prime. If $ab \in (p)$, then $a \in (p)$ or $b \in (p)$.

Proof: By 5.8.2, $(p) = \{np.| n \in Z\}$. Since $ab \in (p)$, we have $ab = kp$ for some $k \in Z$. Therefore $p \mid ab$, and, by 6.2.11, $p \mid a$ or $p \mid b$. Equivalently, $a \in (p)$ or $b \in (p)$.▲

The primeness of p in 6.3.1 suggests the term used to name an ideal that satisfies the converse of superclosure.

6.3.2 DEFINITION. An ideal P in a commutative ring R is called a *prime ideal* if, whenever $ab \in P$ for $a,b \in R$, we have $a \in P$ or $b \in P$.

In Z, (0) and Z are prime ideals. Also, each principal ideal of Z with a prime generator is a prime ideal by 6.3.1. The converse of this result is true in a PID.

6.3.3 THEOREM. Let D be a PID. Every proper prime ideal has a prime generator.

Proof: Let P be a proper prime ideal of D. D is a PID; thus there is some $q \in D$ so that $P = (q)$. Since P is proper, $q \neq 0$ and q is not a unit. (Why?) Therefore q is either prime or composite. Assume, by way of contradiction, that q is composite. Then, by 6.2.10, there exist $a,b \in D$ such that $q = ab$, a and b are not units, and a and b are not associates of q. P is a prime ideal, so $ab \in P$ implies $a \in P$ or $b \in P$. We easily obtain the contradiction now that either a or b is an associate of q. Hence q cannot be a composite and is therefore a prime element of D.▲

We now investigate the structure of the quotient ring associated with a prime ideal.

6.3.4 THEOREM. Let R be a commutative ring with identity e. P is a prime ideal of R if and only if R/P is an integral domain.

Proof: Let P be a prime ideal in R. Clearly R/P is a commutative ring with identity $e + P$. Recall that the zero element in R/P is $P = 0 + P$.

We need to show that R/P has no zero divisors. Assume, by way of contradiction, that $x + P$ is a zero divisor. Then $x + P \neq P$ and there exists $y + P \neq P$ such that $(x + P)(y + P) = P$. But $(x + P)(y + P) = xy + P$ and $xy + P = P$ implies $xy \in P$ (see Exercise 6 in 3.5). Now we use the hypothesis that P is a prime ideal. Since $xy \in P$, we have $x \in P$ or $y \in P$, which in turn implies $x + P = P$ or $y + P = P$. This is our contradiction. Thus there are no zero divisors in R/P and R/P is an integral domain.

Conversely, suppose that R/P is an integral domain. Let $x, y \in R$ so that $xy \in P$. Then $P = xy + P = (x + P)(y + P)$; and since R/P has no zero divisors, we have $x + P = P$ or $y + P = P$. Hence $x \in P$ or $y \in P$ and P is a prime ideal. ▲

The next result is long overdue. Its postponement was due to the fact that it is an elegant corollary to 6.3.4.

6.3.5 COROLLARY. $\langle Z_p, +_p, \cdot_p \rangle$ is a field if and only if $p > 1$ is a prime integer.

Proof: $Z/(p)$ is an integral domain by 6.3.4 and is a field by 6.1.8. $Z_p \approx Z/(p)$ (see Exercise 2 of this section). Thus Z_p is a field. ▲

The second type of ideal we consider is also motivated by the prime-generated ideals of Z. Consider the ideal (3) and suppose that (q) properly contains (3). Then $3 = qn$ for some $n \in Z$; that is, $q \mid 3$. But $q \neq 3$ and $q \mid 3$ implies $q = \pm 1$. Thus $(q) = Z$. So (3) is not properly contained in any ideal except Z itself. In this sense, (3) is a "largest" proper ideal in Z. Every prime-generated ideal of Z possesses this property (see Exercise 3 of this section).

6.3.6 DEFINITION. A proper ideal M in a commutative ring R is a *maximal ideal* of R if the only ideal in which M is properly contained is R itself.

In 6.3.4 we saw that each prime ideal of a commutative ring with identity gives rise to an integral domain. We now show that each maximal ideal of a commutative ring with identity gives rise to a field.

6.3.7 LEMMA. Let R be a commutative ring with identity $e \neq 0$. If R contains no proper ideals, then R is a field.

Proof: As a result of 6.1.6, we need only show that each nonzero element of R has an inverse. Let $a \in R$ with $a \neq 0$. Consider (a). R has no proper ideals and $a \neq 0$; therefore $(a) = R$. Now $e \in (a)$. Thus there exists $x \in R$, so that $e = ax$. Clearly $x = a^{-1}$. ▲

6.3.8 THEOREM. Let R be a commutative ring with identity. M is a maximal ideal of R if and only if R/M is a field.

Proof: From Exercise 6 of Section 5.6 we have the existence of a 1–1 correspondence between the ideals of R that contain M and the ideals of R/M. Suppose, now, that M is maximal. Then, other than R, there are no ideals that contain M properly. By the 1–1 correspondence, R/M contains no proper ideals and, by 6.3.7, R/M is a field.

Conversely, suppose that R/M is a field. Then R/M contains no proper ideals and, by the 1–1 correspondence, R has no ideal that properly contains M except R itself. So M is a maximal ideal of R. ▲

Incorporating the results in 6.3.8, 6.1.7, and 6.3.4, we immediately have the following corollary.

6.3.9 COROLLARY. In a commutative ring with identity, every maximal ideal is a prime ideal.

Proof: Exercise 5 of this section. ▲

Because of 6.3.4 and 6.3.8, we suspect that maximal and prime ideals are not equivalent (since some integral domains are not fields). Thus the converse of 6.3.9 is in doubt. An example of a prime ideal that is not maximal will be given in Section 6.7. If we consider maximal and proper prime ideals in a PID, however, they are one and the same.

6.3.10 THEOREM. Let D be a PID and let I be an ideal of D. I is a proper prime ideal if and only if I is a maximal ideal.

Proof: Let $I = (p)$ be a proper prime ideal of D. By 6.3.3, p is a prime element. Suppose that (q) is such that $(p) \subseteq (q)$. Then $p = sq$ for some $s \in D$; that is, $q \mid p$. Now p is a prime; therefore q is an associate of p or q is a unit. If q is an associate of p, we have $(p) = (q)$ by 6.2.6; if q is a unit, we have $(q) = D$ by 6.2.4. In either case, there is no ideal of D properly containing $(p) = I$ other than D. Hence I is a maximal ideal of D.

The converse is 6.3.9.▲

EXERCISES

†1. Show that (0) in a commutative ring R is prime if and only if R contains no zero divisors.

2. Prove that $Z_p \approx Z/(p)$ for each positive integer p.

3. Show that (p) is a maximal ideal of Z for each prime p without using any of the results of this section.

†4. Give an example of a proper principal ideal in Z that is not maximal.

5. Prove 6.3.9.

†6. Let R be a commutative ring with identity and let $\eta: R \to R'$ be an onto ring homomorphism. Prove:

    (a) R' is an integral domain if and only if ker η is a prime ideal of R, and

    (b) R' is a field if and only if ker η is a maximal ideal of R.

7. Let D be a PID and $p \in D$. Show that the following statements are equivalent.

    (a) p is a prime element of D.

    (b) If $p \mid ab$, then $p \mid a$ or $p \mid b$.

    (c) (p) is a prime ideal of D.

    (d) If $ab \in (p)$, then $a \in (p)$ or $b \in (p)$.

8. Consider $\langle P(Z), +, \cdot \rangle$ (see Example 4 in 5.1 and Exercise 5 in 5.8). $I = \{A \mid A \in P(Z)$ and A is finite$\}$ is an ideal. Prove:

    (a) I is not a prime ideal,

    (b) I is not a maximal ideal in two different ways (*Hint:* Use 6.3.6 and then use the contrapositive of 6.3.9.),

    (c) $(T) = P(T)$ for each $T \in P(Z)$,

    (d) Z is the only element in $P(Z)$ with an inverse,

    (e) every nonzero element is a zero divisor,

    (f) P is a proper prime ideal if and only if P contains every element of Z except one.

6.4 THE FIELD OF QUOTIENTS OF AN INTEGRAL DOMAIN

We have seen that every field is an integral domain. Although the converse of this statement is false, our attempt to "make" a field out of an integral domain has been successful when the system was finite (6.1.8) and when we "divided out" a maximal ideal (6.3.8). Are there other ways of obtaining a field from an integral domain? Our pattern of studying generalizations of the integers suggests that there are. Q is a field of quotients of integers that contains Z. Thus we might expect an integral domain to have a field of quotients. Such is indeed the case.

In this section we show that an integral domain can be imbedded in (or extended to) a field. Since a ring generally is not commutative and contains zero divisors, we cannot expect to extend a ring to a field. Therefore we work with an integral domain. The proof is divided into three parts. First, we will define an equivalence relation and use the equivalence classes for the elements of our field. Second, we will define the field operations on the set of equivalence classes, and, finally, we will display the imbedding isomorphism.

The field we will construct is called the *field of quotients* of an integral domain. Motivation for the name and the construction is supplied by applying this construction to the integers and obtaining the rational numbers—that is, the field of quotients of integers.

6.4.1 THEOREM. An integral domain can be imbedded in a field.

Proof: Let D be an integral domain. If D contains only the zero element, the imbedding is trivial (imbed D in the field with two elements). Suppose, then, that D contains at least two elements.

The equivalence relation. The definition is motivated by equality of fractions. Let $A = \{(x,y) \mid x \in D$ and $y \in D \setminus \{0\}\}$ and define $\sim$ on A by $(a,b) \sim (c,d)$ if $ad = bc$. Exercise 1 of this section shows that $\sim$ is reflexive and symmetric. For transitivity, suppose that $(a,b) \sim (c,d)$ and $(c,d) \sim (x,y)$. We wish to show that $(a,b) \sim (x,y)$ or, equivalently, that $ay = bx$. Now $ad = bc$ and $cy = dx$. Thus $adcy = bcdx$ and $dc(ay - bx) = 0$. D does not have zero divisors, so either $dc = 0$ or $ay - bx = 0$. If $ay - bx = 0$, we immediately have $ay = bx$ and (a,b)

$\sim (x,y)$. If $dc = 0$, then $c = 0$ since $d \in D \setminus \{0\}$. Furthermore, $ad = bc = b0 = 0$ and $d \neq 0$ imply $a = 0$, and $dx = cy = 0y = 0$ and $d \neq 0$ imply $x = 0$. Thus $ay = 0y = b0 = bx$ and $(a,b) \sim (x,y)$. Hence $\sim$ is an equivalence relation on A.

The field. The symbol $\overline{(a,b)}$ is suggested by 2.2.2 for an equivalence class. However, in hopes of gaining an intuitive feeling for the definitions and construction that follow, we will use the symbol $\dfrac{a}{b}$ for the set of all $\sim$ relatives of (a,b); that is,

$$\frac{a}{b} = \{(x,y) \mid (x,y) \in A \quad \text{and} \quad (x,y) \sim (a,b)\}.$$

Notice that $\dfrac{a}{b} = \dfrac{c}{d}$ if and only if $(a,b) \sim (c,d)$.

Let

$$E = \left\{ \frac{a}{b} \,\middle|\, (a,b) \in A \right\}$$

and define $\oplus$ and $\odot$ as follows:

$$\frac{a}{b} \oplus \frac{c}{d} = \frac{ad + bc}{bd}$$

and

$$\frac{a}{b} \odot \frac{c}{d} = \frac{ac}{bd}.$$

Clearly $ad + bc$ and ac are in D; and since D has no zero divisors, we have $bd \neq 0$. Thus $\dfrac{(ad + bc)}{bd}$ and $\dfrac{ac}{bd}$ are elements of E. To show that $\oplus$ and $\odot$ are operations on E, we must show that they are independent of the class representatives. Suppose that $\dfrac{a}{b} = \dfrac{x}{y}$ and $\dfrac{c}{d} = \dfrac{r}{s}$. Then

$$\frac{a}{b} \oplus \frac{c}{d} = \frac{ad + bc}{bd} \quad \text{and} \quad \frac{x}{y} \oplus \frac{r}{s} = \frac{xs + yr}{ys}.$$

Since $ay = bx$ and $cs = dr$, we have

$$(ad + bc)ys = adys + bcys = xsbd + yrbd = (xs + yr)bd$$

and

$$\frac{ad + bc}{bd} = \frac{xs + yr}{ys}.$$

Similarly,

$$\frac{ac}{bd} = \frac{xr}{ys}.$$

Therefore $\oplus$ and $\odot$ are well-defined operations on E. The identities for $\oplus$ and $\odot$ are $\dfrac{0}{b}$ and $\dfrac{b}{b}$ respectively. For $\dfrac{a}{b}$, the additive inverse is $\dfrac{-a}{b}$; the multiplicative inverse is $\dfrac{b}{a}$ provided that $a \neq 0$. Exercise 2 of this section verifies that $\langle E, \oplus, \odot \rangle$ is a field.

The imbedding isomorphism. Let e be the identity in D and let $D' = \left\{ \dfrac{a}{e} \,\middle|\, a \in D \right\}$. Define $\beta: D \to D'$ by $a\beta = \dfrac{a}{e}$ for each $a \in D$. Clearly β is 1–1 and onto. To see that β has the necessary operation-preserving properties, let $a, b \in D$. Then

$$(a + b)\beta = \frac{a + b}{e} = \frac{a}{e} \oplus \frac{b}{e} = a\beta \oplus b\beta$$

and

$$(ab)\beta = \frac{ab}{e} = \frac{a}{e} \odot \frac{b}{e} = a\beta \odot b\beta.$$

Therefore β is a ring isomorphism and the field E is an extension of D. $\blacktriangle$

6.4.2 COROLLARY. Let R and S be integral domains. If R and S are isomorphic, then their fields of quotients are isomorphic.

Proof: Exercise 6 of this section. $\blacktriangle$

This corollary tells us that the field of quotients of an integral domain is unique up to isomorphism.

EXERCISES

Exercises 1 through 5 refer to the proof of 6.4.1.

1. Show that $\sim$ is reflexive and symmetric on A.
2. Verify that $\langle E, \oplus, \odot \rangle$ is a field.
†3. Where did we use the fact that D has at least two elements?
4. Let b be fixed in $D \setminus \{0\}$. If we had defined $D' = \{a/b \mid a \in D\}$ and $\beta: D \to D'$ by $a\beta = a/b$ for $a \in D$, would β still be a ring isomorphism? Justify your answer.
5. Let b be fixed in $D \setminus \{0\}$. If we had defined $D' = \{ab/b \mid a \in D\}$ and $\beta: D \to D'$ by $a\beta = ab/b$ for $a \in D$, show that β would still be a ring isomorphism. This tells us that we did not need to require D to have an

identity in order to imbed it in a field—that is, taking β to be this map shows that a commutative ring without zero divisors has a field of quotients.

6. Prove 6.4.2.

7. By Exercise 5, a commutative ring without zero divisors has a field of quotients. If R and S are isomorphic commutative rings without zero divisors, show that their fields of quotients are isomorphic.

†8. Construct the field of quotients of the ring of even integers (by Exercise 5, a commutative ring without zero divisors has a field of quotients). Show that this field is isomorphic to Q.

†9. Construct a counterexample to the converse of Exercise 7.

10. Let T be a field. By 6.1.7, T is an integral domain and therefore has a field of quotients. Show that T is isomorphic to its field of quotients.

11. Let R and R' be isomorphic rings. If a is a zero divisor in R, show that a', the copy of a in R', is a zero divisor in R'.

†12. Why can't Z_n, for n composite, be imbedded in a field? (*Hint:* Use Exercise 11.)

13. (Refer to the proof of 6.4.1.) Let T be a field that contains D. Show that E can be imbedded in T. In this sense, E is the smallest field extension of D. (*Hint:* Consider $\eta: E \to T$ defined by $(a/b)\ \eta = ab^{-1}$.)

6.5 ORDER IN AN INTEGRAL DOMAIN

We have used the order relation "is less than" on the integers from time to time. For instance, order played a big part in the development of the division algorithm and in showing that every subgroup of a cyclic group is cyclic. The generalization of this concept to an integral domain so that the familiar properties are retained is the goal of this section.

All of us are aware that positive integers are "those integers that are greater than zero." However, how would we actually show $1 > 0$ in order to conclude that 1 is a positive integer? We answer this question in the more general setting of an integral domain.

6.5.1 DEFINITION. A nonempty subset D_+ of an integral domain D is a *positive set* if D_+ satisfies the following conditions:

1. for each $a,b \in D_+$, we have $(a + b) \in D_+$ and $ab \in D_+$, and
2. for each $a \in D$, exactly one of $a \in D_+$, $-a \in D_+$, or $a = 0$ holds.

The elements of D_+ are called *positive elements* and the nonzero elements of D that are not in D_+ are called *negative elements*.

Since D_+ satisfies (2) and is nonempty, an integral domain must have at least two elements to have a positive set. We will actually show in 6.5.8 that an integral domain must be infinite in order to have a positive set.

6.5.2 DEFINITION. An integral domain is an *ordered integral domain* if it contains a positive set.

Obviously Z is an ordered integral domain, since Z_+ is a positive set. Later we will show that Z_+ is the only subset of Z which can be a positive set. Q and F are also ordered integral domains.

In order to extend the natural concept that b "is less than" a if $a - b$ is "positive," we use a positive set to induce the relation $<$ on an ordered integral domain.

6.5.3 DEFINITION. Let D_+ be a positive set of an ordered integral domain D. For $a, b \in D$, b is *less than* a, denoted $b < a$, if $(a - b) \in D_+$. Equivalent notation and terminology are as follows: a is *greater than* b, denoted $a > b$, if $(a - b) \in D_+$.

In Z, $(-3) - (-5) = 2 \in Z_+$. Therefore $-5 < -3$ and $-3 > -5$. Also, $(1 - 0) \in Z_+$. Thus $1 > 0$ and 1 is a positive element. In general, $a > 0$ if and only if $a \in D_+$. Using this fact, we can restate the properties of D_+ in 6.5.1 as follows:

1. for $a > 0$ and $b > 0$, we have $a + b > 0$ and $ab > 0$, and
2. for $a \in D$, exactly one of $a > 0$, $a < 0$, or $a = 0$ holds.

Using the restatement of 6.5.1 and appealing to the familiar properties of $<$ on Z, we have the following theorem.

6.5.4 THEOREM. Let D be an ordered integral domain and $a, b, c \in D$. Then
1. $<$ satisfies the *trichotomy property*; that is, exactly one of $a > b$, $a < b$, or $a = b$ holds,

2. $a > 0$ if and only if a is a positive element, and

3. $<$ is transitive; that is, if $a < b$ and $b < c$, then $a < c$.

Proof: 1. Let D_+ be the positive set that induced the order on D. Consider the element $(a - b) \in D$. By 6.5.1.2, exactly one of $(a - b) \in D_+$, $(b - a) \in D_+$, or $a - b = 0$. Therefore exactly one of $a > b$, $a < b$, or $a = b$ holds.

Exercise 1 of this section establishes points (2) and (3). ▲

Again we appeal to the integers for motivation. The popular mnemonic phrase "unequals added to unequals remains unequal in the same sense" is still valid, and the fact that positive (negative) elements preserve (reverse) inequalities via multiplication also holds.

6.5.5 THEOREM. Let D be an ordered integral domain and $a,b,c,d \in D$.

1. If $a < b$, then $a + c < b + c$.
2. If $a < b$ and $c < d$, then $a + c < b + d$.
3. If $a < b$ and $c > 0$, then $ac < bc$.
4. If $a < b$ and $c < 0$, then $ac > bc$.

Proof: Exercise 2 of this section. ▲

We have already mentioned that $1 > 0$ since $1 \in Z_+$. However, this only tells us $1 > 0$ as long as we are using Z_+ for the positive set in Z. It is possible that there could exist a positive set in Z different from Z_+ that does not contain 1. That this is not the case is immediate from the following corollary.

6.5.6 COROLLARY. If D is an ordered integral domain and $0 \neq a \in D$, then $a^2 > 0$.

Proof: Either $a > 0$ or $a < 0$. If $a > 0$, then $a^2 > 0$ by 6.5.5.3. If $a < 0$, then $a^2 > 0$ by 6.5.5.4. ▲

If e is the identity of an ordered integral domain D, clearly $e \neq 0$. Thus $e = e^2 > 0$ by 6.5.6. Hence $1 > 0$ no matter what positive set of Z is used to induce $<$. Using this fact, we can conclude that Z_+ is the only positive set of Z (Exercise 5 of this section).

From the study of mappings that preserve operations, we have obtained valuable information about groups and rings. We end this section with a look at mappings that in addition to their normal operation-preserving duties, also preserve order.

6.5.7 DEFINITION. Let D and D' be ordered integral domains and $\alpha: D \to D'$ be an isomorphism. If $a < b$ in D implies $a\alpha < b\alpha$ in D', then α is called an *order isomorphism*.

The next two results make use of an order isomorphism. The first characterizes the integers as the smallest ordered integral domain (up to isomorphism). The second shows that the field of quotients of an ordered integral domain D has a natural order induced by D and preserved by the imbedding isomorphism.

6.5.8 THEOREM. If D is an ordered integral domain, then Z can be imbedded in D via an order isomorphism.

Proof: Let e be the identity in D ($e \neq 0$ since D is an ordered integral domain). Consider the cyclic additive subgroup $\langle e \rangle$. Note that $\langle e \rangle$ is different from (e), since $\langle e \rangle = \{ne \mid n \in Z\}$ and $(e) = \{de \mid d \in D\}$. Exercise 6 of this section verifies that $\langle e \rangle$ is an ordered subdomain of D—that is, an ordered integral domain in its own right. Define $\alpha: Z \to \langle e \rangle$ by $n\alpha = ne$ for all $n \in Z$. Exercise 8 of this section establishes that α is an order isomorphism. Hence D is an extension of Z.▲

6.5.9 THEOREM. An ordered integral domain can be imbedded in an ordered field via an order isomorphism.

Proof: Let D be an ordered integral domain and let E be the corresponding field of quotients. We intend to induce order on E by using the order of D. Then we will show that the imbedding of D in E is order preserving.

Ordering E. Recall that $E = \{a/b \mid a \in D$ and $b \in D \setminus \{0\}\}$. Let $E_+ = \{a/b \mid a/b \in E$ and $ab > 0$ in $D\}$. To ensure that E_+ is well defined, we need to show that the elements are independent of their representatives. Let $a/b \in E_+$ and suppose that $a/b = c/d$; then $ad = bc$. Now $ab > 0$, so $a < 0$ and $b < 0$ or $a > 0$ and $b > 0$. In either case, since ad

$= bc$ we must have $cd > 0$. Therefore $c/d \in E_+$ and E_+ is well defined. Exercise 9 of this section shows that E_+ is a positive set. Hence E is an ordered integral domain and, of course, an ordered field.

Imbedding D in E. In the proof of 6.4.1, $\beta: D \to D'$, defined by $a\beta = a/e$ for all $a \in D$, was seen to be a ring isomorphism. We complete the proof of this theorem by showing that β preserves order. Exercise 10 of this section verifies that $a/e > b/e$ in E if and only if $a > b$ in D. Let $a > b$ in D. Then $a\beta = a/e > b/e = b\beta$; that is, $a\beta > b\beta$.▲

Since every ordered field is an ordered integral domain, it now follows that the rationals form the smallest ordered field in the same sense that the integers form the smallest ordered integral domain.

6.5.10 COROLLARY. If E is an ordered field, then Q can be imbedded in E via an order isomorphism.

Proof: Exercise 11 of this section.▲

EXERCISES

1. Prove 6.5.4.2 and 6.5.4.3.
2. Prove 6.5.5.
†3. Let D be an ordered integral domain and $a,b,c \in D$. Prove:
    (a) if $a > e$, then $a^2 > a$,
    (b) if $a > 0$ and $ab > ac$, then $b > c$,
    (c) if $a > b$, then $-a < -b$,
    (d) if a^{-1} exists and $a > 0$, then $a^{-1} > 0$.
4. Let D be an ordered integral domain and define
    $$|a| = \begin{cases} a \text{ if } a > 0 \\ 0 \text{ if } a = 0 \\ -a \text{ if } a < 0 \end{cases} \quad \text{for all } a \in D.$$
    Prove:
    (a) if $a \neq 0$, then $|a| > 0$,
    (b) $|ab| = |a||b|$,
    (c) $-|a| \leq a \leq |a|$ ($\leq$ refers to less than or equal to),
    (d) $|a + b| \leq |a| + |b|$.
5. Show that Z_+ is the only positive set of Z.
†6. Show that $\langle e \rangle$ is an ordered subdomain of an ordered integral domain D with identity e.
†7. Show that an ordered integral domain cannot have a largest element. Can a finite integral domain be ordered in the sense of 6.5.2?

8. Establish that α, as defined in the proof of 6.5.8, is an order isomorphism.
9. Show that E_+, as defined in the proof of 6.5.9, is a positive set of E.
†10. Show that $a/e > b/e$ in E if and only if $a > b$ in D, where D and E are as in the proof of 6.5.9.
11. Prove 6.5.10.

6.6   THE CHARACTERISTIC

The additive order of an element is studied in both group and ring theory. In ring theory, however, we are more interested in a property similar to 3.7.10a in Section 4.5 (let G be an additive group with identity 0 and $|G| = n$; then $ng = 0$ for all $g \in G$) than in the additive order of individual elements. In the ring of integers, each nonzero element has infinite additive order; but in the ring Z_n, the order of each nonzero element is positive and, in fact, is a divisor of n. This property is a valuable distinction between these rings and leads us to what is called the characteristic of a ring.

6.6.1 DEFINITION. If R is a ring with the property that for some positive integer n we have $nx = 0$ for all $x \in R$, then the smallest positive integer for which this is true is called the *characteristic* of R. We write Char $R = n$ if such a smallest positive integer n exists. If no such positive integer exists, R is said to have *characteristic zero* (since $nx = 0$ for all $x \in R$ implies $n = 0$).

Note 1. If any positive integer n has the property that $nx = 0$ for all $x \in R$, then a smallest such positive integer exists because each nonempty set of positive integers has a smallest element (i.e., because Z_+ is well ordered).

Note 2. Recall that nx is a multiple of x and means x added to itself n times. The basic properties of multiples are: $nx = xn$, $(nx)y = n(xy) = x(ny)$, $m(nx) = (mn)x = n(mx)$, $(n + m)x = nx + mx$, and $n(x + y) = nx + ny$ for all $m, n \in Z$ and $x, y \in R$.

The distinction between rings of positive characteristic and rings of characteristic zero is valuable enough to cause the two cases to be treated separately and is especially fruitful in field theory. We proceed up the ladder from rings to integral domains and fields by first considering a ring with identity.

6.6.2 THEOREM. Let R be a ring with identity e. Char R $= n > 0$ if and only if n is the smallest positive integer so that $ne = 0$.

Proof: Suppose that Char $R = n > 0$; then clearly $ne = 0$. Assume, by way of contradiction, that there is an $r \in Z$ with $0 < r < n$ such that $re = 0$. Then for all $x \in R$ we have $rx = r(ex) = (re)x = 0x$ $= 0$; therefore r has the same property as n and $r < n$. This is the desired contradiction. Thus n is the smallest positive integer such that $ne = 0$.

Conversely, let n be the smallest positive integer such that $ne = 0$. Then $n \leq$ Char R (why?). Now, for all $x \in R$, $nx = n(ex)$ $= (ne)x = 0$. Thus Char $R \leq n$ and equality holds.▲

This theorem enables us to find the characteristic of a ring with identity by examining the additive order of just one element instead of having to check the orders of all elements, as the definition seems to require. We also obtain the immediate corollary that the characteristic for integral domains and fields is the additive order of their multiplicative identity when this order is finite and zero otherwise. Upon examining the characteristic of an integral domain more closely, we have the following theorem.

6.6.3 THEOREM. Let D be an integral domain with at least two elements. Then Char $D = 0$ or Char $D = p$ for some prime p.

Proof: Let Char $D = n > 0$ and suppose that n is composite. Then there exist $r, s \in Z_+$ such that $r < n$, $s < n$, and $rs = n$. From 6.6.2, n is the smallest positive integer such that $ne = 0$. Thus $re \neq 0$ and $se \neq 0$. Now $(re)(se) = (rs)e = ne = 0$ and re is a zero divisor. This result is a contradiction, since D is an integral domain. Therefore n is not composite. Clearly $n > 1$ and is not a unit in Z. Hence n is a prime.▲

6.6.4 COROLLARY. The characteristic of a field is either zero or a prime.

Proof: Exercise 3 of this section.▲

6.6.5 COROLLARY. In an integral domain with a positive

characteristic, every nonzero element has additive order equal to the characteristic.

Proof: Exercise 4 of this section. ▲

Utilizing the characteristic enables us to show the close relationship between Z and an integral domain with characteristic zero and between Z_p and an integral domain of characteristic p.

6.6.6 THEOREM. An integral domain with characteristic p is an extension of Z_p, and an integral domain with characteristic zero is an extension of Z.

Proof: Let D be an integral domain with identity e. Define $\alpha: Z \to D$ by $n\alpha = ne$ for all $n \in Z$. Exercise 5 of this section verifies that α is a ring homomorphism.

Case 1. Let Char $D = p$, p a prime. Then our claim is ker α $= (p)$. Clearly $(p) \subseteq$ ker α. To establish equality, let $m \in$ ker α. Then $m\alpha = me = 0$, and since Char $D = p$, we have $p \mid m$; that is, $m \in (p)$. Therefore ker $\alpha \subseteq (p)$ and equality holds. By 5.6.4, $Z\alpha \approx Z/(p)$ and, by Exercise 2 of 6.3, $Z/(p) \approx Z_p$. The relation $\approx$ is transitive; thus $Z_p \approx Z\alpha$. Finally, we have Z_p imbedded in D, since $Z\alpha \subseteq D$.

Case 2. Suppose, now, that Char $D = 0$. Then α is 1-1 (Exercise 6 of this section). Therefore $Z \approx \{ne \mid n \in Z\}$; and since $\{ne \mid n \in Z\} \subseteq D$, we have Z imbedded in D. ▲

EXERCISES

1. Let R be a ring. Show that $(m + n)x = mx + nx$ and $m(x + y)$ $= mx + my$ for all $m, n \in Z$ and $x, y \in R$.
†2. Prove that Char $Z_n = n$ for every positive integer n.
3. Prove 6.6.4.
4. Prove 6.6.5.
5. Show that α, as defined in the proof of 6.6.6, is a ring homomorphism.
6. Show that α, as defined in the proof of 6.6.6, is 1-1 when Char $D = 0$.
†7. Let D be an integral domain with nine elements. Prove that Char $D = 3$. (*Hint:* Use 6.6.6.)
8. Show that Z_p, p a prime, has no proper subfield.

†9. Show that Q does not contain a proper subfield.

†10. Char $Z_6 = 6$. Is there an element in Z_6 whose additive order is less than 6?

11. Prove that if R is a ring with no divisors of zero, then Char $R = 0$ or Char $R = p$, p a prime.

12. If E is a field with characteristic p, prove that E is an extension of Z_p.

13. If E is a field with characteristic zero, prove that E is an extension of Z.

14. Show that $\langle P(X), +, \cdot \rangle$ has characteristic 2 (see Example 4 in 5.1).

6.7 POLYNOMIALS

Polynomials hold an esteemed position in mathematics. They are studied extensively from high school algebra on through advanced abstract algebra that is far beyond the scope of this book. We present only the grass roots here, just enough to motivate further study and exemplify some concepts that have been deferred to this section.

You are undoubtedly familiar with expressions like $x^2 + 5x + 6$, $3x - 7$, and $x^2 - 4$. Such "polynomials" were probably studied as parts of equations like $x^2 + 5x + 6 = 0$, $3x - 7 = -2$, and $x^2 - 4 = -1$, where solutions were sought. The polynomial equation $x^2 + 5x + 6 = 0$ is easily solved by writing $(x + 3)(x + 2) = 0$ from which we conclude that $x = -3$ or -2 (why? no zero divisors). Notice that the coefficients in $x^2 + 5x + 6 = 0$ and the solutions are integers. This is not the case for $3x - 7 = -2$ and $x^2 - 4 = -1$. For $3x - 7 = -2$, the solution is $x = \frac{5}{3}$, a rational number; whereas for $x^2 - 4 = -1$, the solutions $x = \pm\sqrt{3}$ are irrational numbers. The coefficients of powers of x appear to be restricted to the integers; yet the values of x vary from the integers to the rationals to the reals. Permissible values for the symbol x can be in a system that is an extension of the system containing the coefficients. For this reason, the symbol x is called an *indeterminant* over the system of coefficients.

Our formal definition of a polynomial is by no means the most general one that can be given, but it is quite suitable for establishing the goals of this section.

6.7.1 DEFINITION. Let R be a commutative ring with identity e, $x \in R$, and S be a subring of R containing e. A *polynomial* in x over S is a ring element of R having the form

$$\alpha = a_0 + a_1 x + a_2 x^2 + \cdots + a_n x^n,$$

where n is a positive integer and $a_i \in S$ for $i = 0, 1, 2, \ldots, n$. The elements $a_i \in S$ are called the *coefficients* of α, and the symbol $S[x]$ is used to denote the collection of all polynomials in x over S.

Since polynomials in $S[x]$ are elements of the ring R, they can be added and multiplied via the operations in R. It is important to note that $0x^k = 0$ (both 0's in R) so that the symbols $a_0 + a_1x + a_2x^2$ and $a_0 + a_1x + a_2x^2 + 0x^3 + 0x^4$ represent the same polynomial in $S[x]$. We shall agree to omit or include terms of the form $0x^k$, depending on which is more desirable. We can thus write $x^2 + 3$ for $x^2 + 0x + 3$ in $Z[x]$ (note that the coefficient of x^2 is the identity $1 \in Z$). With this convention, polynomials in $S[x]$ can be considered more elegantly in the form $a_0 + a_1x + a_2x^2 + \cdots$, where only a finite number of the coefficients are not the zero element in S. This convention also offers easily obtained formulas for addition and multiplication.

Addition can be accomplished by simply adding the coefficients of "like powers" of x in the respective addends to obtain the coefficient of that particular power of x in the sum. For example,
$$(a_0 + a_1x + a_2x^2) + (b_0 + b_1x)$$
$$= (a_0 + b_0) + (a_1x + b_1x) + (a_2x^2 + 0x^2) \qquad \text{(Why?)}$$
$$= (a_0 + b_0) + (a_1 + b_1)x + (a_2 + 0)x^2. \qquad \text{(Why?)}$$
In the more general situation where $\alpha = a_0 + a_1x + a_2x^2 + \cdots$ and $\beta = b_0 + b_1x + b_2x^2 + \cdots$, we have
$$\alpha + \beta = (a_0 + b_0) + (a_1 + b_1)x + (a_2 + b_2)x^2 + \cdots$$
where only a finite number of the coefficients are different from the zero in S.

Notationally, multiplication is more complicated than addition, since obtaining the coefficient of x^k requires consideration of all terms of the form $a_rx^rb_sx^s = a_rb_sx^{r+s}$, where $r + s = k$. We exemplify and then generalize via the following:
$$(a_0 + a_1x)(b_0 + b_1x + b_2x^2)$$
$$= a_0(b_0 + b_1x + b_2x^2) + a_1x(b_0 + b_1x + b_2x^2)$$
$$= a_0b_0 + (a_0b_1x + a_1b_0x) + (a_0b_2x^2 + a_1b_1x^2) + a_1b_2x^3$$
$$= a_0b_0 + (a_0b_1 + a_1b_0)x + (a_0b_2 + a_1b_1)x^2 + a_1b_2x^3.$$
Again considering the general α and β in $S[x]$, we obtain
$$\alpha + \beta = a_0b_0 + (a_0b_1 + a_1b_0)x + (a_0b_2 + a_1b_1 + a_0b_2)x^2$$
$$+ (a_0b_3 + a_1b_2 + a_2b_1 + a_3b_0)x^3 + \cdots$$
In our first theorem about polynomials, we examine the structure of $S[x]$.

6.7.2 THEOREM. If R is a commutative ring with identity e, $x \in R$, and S is a subring containing e, then $S[x]$ is a commutative ring with identity.

Proof: Let $\alpha = a_0 + a_1 x + \cdots$ and $\beta = b_0 + b_1 x + \cdots$ be polynomials in x over S. Clearly
$$-\beta = -e(b_0 + b_1 x + \cdots) = -b_0 - b_1 x - \cdots,$$
so $\alpha - \beta = (a_0 - b_0) + (a_1 - b_1)x + \cdots$ is in $S[x]$. Via the formula for multiplication, $\alpha\beta \in S[x]$; thus, by 5.5.2, $S[x]$ is a subring of R and hence is a ring. $S[x]$ inherits commutativity from R and contains e since $S \subseteq S[x]$. ▲

If $x \in S$, then $S = S[x]$; if $x \notin S$, then $S \subset S[x]$ (see Exercises 2 and 3 of this section). Therefore $S[x]$ is truly an extension of S if $x \notin S$, and we can justifiably call $S[x]$ the *ring of polynomials in x over S*. We adopt the convention of referring to $S[x]$ in this way when S is a commutative ring with identity rather than referring each time to the super ring R which contains S as a subring. This convention makes future reference to polynomial rings much more convenient, especially in the statement of definitions, theorems, and examples.

The following definition formalizes some concepts with which we are familiar.

6.7.3 DEFINITION. Let S be a commutative ring with identity and let $\alpha = a_0 + a_1 x + \cdots$ be in $S[x]$. If $\alpha \neq 0$, then the largest non-negative integer n such that $a_n \neq 0$ is called the *degree* of α and we write $\deg(\alpha) = n$. The elements a_n and a_0 are called the *leading coefficient* and *constant term* of α respectively. If α is the zero polynomial (i.e., $a_i = 0$ for all i), then α has no degree.

The relationship of the degrees of α and β to the degrees of $\alpha + \beta$ and $\alpha\beta$ is partially given in the following

6.7.4 THEOREM. Let S be a commutative ring with identity. If $\alpha, \beta \in S[x]$, $\alpha \neq 0$, $\beta \neq 0$, $\alpha + \beta \neq 0$, and $\alpha\beta \neq 0$; then
   1. $\deg(\alpha + \beta) \leq$ maximum $(\deg \alpha, \deg \beta)$, and
   2. $\deg(\alpha\beta) \leq \deg(\alpha) + \deg(\beta)$.

Proof: Exercise 4 of this section. ▲

Part (2) of 6.7.4 can be strengthened to equality if S is an integral domain because of the absence of zero divisors. This fact is necessary in an upcoming major result; therefore we present it here as a lemma.

6.7.5 LEMMA. Let D be an integral domain and let $\alpha, \beta \in D[x] \setminus \{0\}$. Then

$$\deg(\alpha\beta) = \deg(\alpha) + \deg(\beta).$$

Proof: Let $\deg(\alpha) = m$ and $\deg(\beta) = n$. Since neither α nor β is the zero polynomial, we have that their leading coefficients are not zero—that is, $a_m \neq 0$ and $b_n \neq 0$. D is an integral domain; thus $a_m b_n \neq 0$ in D. Now $a_m b_n$ is the coefficient of x^{m+n} in $\alpha\beta$, so $\deg(\alpha\beta) \geq m + n = \deg(\alpha) + \deg(\beta)$. Equality now follows from 6.7.4.▲

Many questions arise concerning the structure of polynomial rings. What relation does the structure of S have to the structure of $S[x]$? If a theorem is valid in S, does it still hold in $S[x]$? We have already seen that with simply the minimum requirement that S be a commutative ring with identity, $S[x]$ is also a commutative ring with identity. Next we examine the relationship between S and $S[x]$ when one is an integral domain.

6.7.6 THEOREM. Let D be a commutative ring with identity. $D[x]$ is an integral domain if and only if D is an integral domain.

Proof: Suppose that $D[x]$ is an integral domain. Then $D \subseteq D[x]$ and D is a commutative ring with identity—hence an integral domain.

Conversely, suppose that D is an integral domain. By 6.7.2, $D[x]$ is a commutative ring with identity so that we need only establish the absence of zero divisors in $D[x]$. Let $\alpha, \beta \in D[x] \setminus \{0\}$.

Case 1. $\deg(\alpha) > 0$ or $\deg(\beta) > 0$. By 6.7.5, $\deg(\alpha\beta) = \deg(\alpha) + \deg(\beta) > 0$, so $\alpha\beta \neq 0$.

Case 2. Suppose that $\deg(\alpha) = \deg(\beta) = 0$. Then $\alpha = a_0 \neq 0$ and $\beta = b_0 \neq 0$, so $\alpha\beta = a_0 b_0 \neq 0$. Thus neither α nor β is a zero divisor and $D[x]$ is an integral domain.▲

Having now developed enough machinery to consider our concluding examples, we leave further pursuit of polynomials to the interested reader. The literature abounds with excellent works in this area.

EXAMPLES

1. An integral domain that is not a PID.

Consider F as a ring with identity 1 and Z as a subring of F. Let $x \in F \setminus Z$. Now $Z[x]$ is an integral domain because Z is, and x is a prime element of $Z[x]$. Also, all prime elements of Z are prime elements of $Z[x]$. Therefore 3 and x are prime elements of $Z[x]$. Let I denote the ideal of $Z[x]$ generated by $\{3,x\}$—that is, $I = (3,x)$. Suppose, by way of contradiction, that $Z[x]$ is a PID. Then there exists $\beta \in Z[x]$ such that $I = (\beta)$. Now $(\beta) = \{\alpha\beta \mid \alpha \in Z[x]\}$ and $\{3,x\} \subset (\beta)$; thus $\beta \mid 3$ and $\beta \mid x$. As distinct primes in $Z[x]$, the only common divisors of 3 and x are units, so β is a unit and $I = (\beta) = Z[x]$. This result is our desired contradiction, since $I = \{\alpha 3 + \gamma x \mid \alpha, \gamma \in Z[x]\}$ and thus does not contain any elements of $Z[x]$ with a constant term different from a multiple of 3 (i.e., $I \neq Z[x]$). Therefore $Z[x]$ is not a PID.

This example also answers another structural question by showing that $D[x]$ is not necessarily a PID even though D is a PID.

2. A prime ideal that is not a maximal ideal.

Consider (x) in $Z[x]$, where $x \in F \setminus Z$ as in Example 1. Clearly $(x) \subset (3,x) \subset Z[x]$; therefore (x) is not a maximal ideal. It remains to show that (x) is a prime ideal. First note that $(x) = \{\alpha x \mid \alpha \in Z[x]\}$ $= \{\gamma \mid \gamma \in Z[x]$ and that γ has zero as its constant term$\}$. Suppose that for $\alpha, \beta \in Z[x]$, we have $\alpha\beta \in (x)$. Then the constant term of $\alpha\beta$ is zero—that is, $a_0 b_0 = 0$—where a_0 and b_0 are the constant terms of α and β respectively. Z contains no zero divisors; therefore $a_0 = 0$ or $b_0 = 0$ and, equivalently, $\alpha \in (x)$ or $\beta \in (x)$. Hence (x) is a prime ideal.

EXERCISES

1. Determine $\alpha + \beta + \gamma$ and $\alpha\beta\gamma$ when $\alpha = 2 + x$, $\beta = x + 3x^2$, and $\gamma = 1 + 3x + 2x^2$.

†2. Show that $Z[2] = Z$ and $Z \subset Z[\frac{1}{2}]$.

3. Let R be a commutative ring with identity e and let S be a subring of R containing e. Show that $S = S[x]$ if $x \in S$ and $S \subset S[x]$ if $x \in R \setminus S$.

4. Prove 6.7.4.

†5. Exhibit a choice for $\alpha, \beta \in Z[x]$ so that $\deg(\alpha + \beta) < \text{maximum}$ $[\deg(\alpha), \deg(\beta)]$.

†6. Consider $Z_6[x]$. Exhibit a choice for $\alpha, \beta \in Z_6[x]$ such that $\deg(\alpha \cdot_6 \beta)$ $< \deg(\alpha) + \deg(\beta)$. (*Hint:* Make use of the zero divisors in Z_6.)

 7. Let S be a commutative ring with identity. Prove that the set A of all polynomials of $S[x]$ with the property that all odd powers of x have zero coefficients is a subring. Also, show that A is not an ideal.

 8. Show that each prime element in Z is also a prime element in $Z[x]$. Then show, with x as in Example 1, that x is a prime element of $Z[x]$. (*Hint:* In each case, use 6.7.5.)

 9. Clearly $(x) \subseteq (3, x)$ for x as in Example 1. Justify $(x) \subset (3, x)$.

†10. Show that $Z[x]$ is an ordered integral domain by showing that $A = \{\alpha \mid \alpha \in Z[x]$ and the leading coefficient of α is a positive integer$\}$ is a positive set in $Z[x]$.

ANSWERS TO SELECTED EXERCISES

CHAPTER 1

Section 1

2. (a) $\backslash q \to p$.　　(b) $p \wedge q$.　　(c) $p \leftrightarrow q$.

5.

p	q	$p \to q$	$q \to p$	$(p \to q) \wedge (q \to p)$	$p \leftrightarrow q$	$[(p \to q) \wedge (q \to p)]$ $\leftrightarrow (p \leftrightarrow q)$
T	T	T	T	T	T	T
T	F	F	T	F	F	T
F	T	T	F	F	F	T
F	F	T	T	T	T	T

6. $(p \lor q) \land \setminus (p \land q)$.

Section 2

3. $2 + 1 = 4$ is *false*; therefore the implications "if $2 + 1 = 4$, then $3 - 4 = 0$" and "if $2 + 1 = 4$, then $2 + 1 \neq 5$" are true by 1.1.1.

7. The error is assuming that the converse is true.

8. $p \leftrightarrow q$ is true. This suggests that a proof for $p \rightarrow q$, together with a proof for $q \rightarrow p$, suffices as a proof for $p \leftrightarrow q$.

Section 3

2. (a) $q \rightarrow p, \setminus q \rightarrow \setminus p$.   (b) $p \rightarrow q; \setminus q \rightarrow \setminus p$.

    (c) If lines L_1 and L_2 are parallel, then they have no points in common. If lines L_1 and L_2 are not parallel, then they have points in common.

7.

p	q	$p \rightarrow q$	$\setminus (p \land \setminus q)$	$\setminus (p \land \setminus q) \leftrightarrow (p \rightarrow q)$
T	T	T	T	T
T	F	F	F	T
F	T	T	T	T
F	F	T	T	T

8. $p \rightarrow q$ is the negation of $p \land \setminus q$.

Section 4

2. Assume that $p \land \setminus q$ is true; that is, let xy be even and let y be odd. Then there exist integers m and n such that $xy = 2m$ and $y = 2n + 1$. Let $x = 2x + t$, where $t = 0$ or 1 (if $t = 0$, then x is even and if $t = 1$, then x is odd). Now $2m = xy = (2s + t)(2n + 1) = 4sn + 2nt + 2s + t = 2(2sn + nt + s) + t$. Therefore $t = 0$ and x is even. Hence, by 1.4.1, the proof is complete.

7. (a) Let $x = 4$. Then x is even and x^2 is even.

    (b) Let $x = 3$. Then x^2 is 9 and $x \neq -3$.

    (c) Let $x = 0$. Then $x^2 + 4 = (x + 2)^2$.

Section 5

3. (a) True   (b) False   (c) True   (d) False

5. Suppose that there exists a set A such that $\emptyset \nsubseteq A$. Then there is an element x for which $x \in \emptyset$ and $x \notin A$. Clearly this is a contradiction, for $\emptyset$ has no elements. Therefore for every set A we have $\emptyset \subseteq A$.

7. Assume that $A \subseteq B$ and $B \subseteq C$. Let $x \in A$. Then $x \in B$, since $A \subseteq B$. Similarly, $x \in C$, since $x \in B$. Now we have that every element of A is also an element of C; that is, $A \subseteq C$.

Section 6

2. (a) $\{6,7,8\}$ (b) $\{6,7,8\}$
    (c) $\{1,2,3,4,6,7,8,9,10\}$ (d) $\{1,2,3,4,6,7,8,9,10\}$

6. (a) $A = \emptyset$.   (b) $A = \emptyset$.   (c) $A \subseteq B$.   (d) $B \subseteq A$.
    (e) $A \cap B = \emptyset$.   (f) $A \subseteq B$.   (g) $A = B = \emptyset$.   (h) $A = U$.

Section 7

4. (c) Let $A \cup B = C$ and $A \cap B = \emptyset$. We wish to show that $A = C \setminus B$.
    Let $x \in A$. Then $x \in C$, since $A \subseteq C$; also, $x \in B$, since $A \cap B = \emptyset$.
    Thus $x \in C \setminus B$ and $A \subseteq C \setminus B$. Let $y \in C \setminus B$. Then $y \in C$ and $y \notin B$.
    Now $y \in A \cup B$ and $y \notin B$ imply $y \in A$. Hence $C \setminus B \subseteq A$, and equality
    holds.

6. Since $0 \in M_n$ for each $n \in Z_+$, we have $0 \in \cap \mathscr{F}$. Now let $x \in F$ with
    $x \neq 0$. Then there is a positive integer r such that $|x| > 1/r$. Now
    $x < -1/r$ or $x > 1/r$; thus $x \notin M_r$ and $x \notin \cap \mathscr{F}$. Hence $\cap \mathscr{F} = \{0\}$.

CHAPTER 2

Section 1

2. (a)   (b)

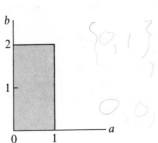 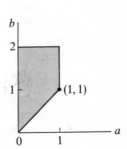

5. (a) r, s, t (b) $\backslash r, s, \backslash t$ (c) r, s, t (d) $\backslash r, \backslash s, t$

Section 2

2. (a) Must add $(3,3)$ and $(3,2)$.
    (b) $\mathscr{P} = \{\{1\}, \{2,3\}\}$.

4. An equivalence class with a unique representative has only one element.
    Equality on Z is such that every equivalence class has exactly one element.

Section 3

3. Let α and β be maps. Suppose that $\alpha = \beta = A \times B$ in the sense of set
    equality. Then dom $\alpha \times A\alpha = $ dom $\beta \times A\beta$ and dom $\alpha = $ dom β.

Furthermore, $(a,a\alpha) = (a,a\beta)$ for all $a \in A$; therefore $a\alpha = a\beta$ for all $a \in A$.

Conversely, suppose that dom $\alpha = $ dom $\beta = A$ and $a\alpha = a\beta$ for all $a \in A$. Then $A\alpha = A\beta$ and $\alpha = $ dom $\alpha \times A\alpha = $ dom $\beta \times A\beta = \beta$.

7. We show that $\alpha\alpha^{-1} = i_A$. α is 1-1 and onto; thus α^{-1} exists. For each $a \in A$, $(a,a\alpha) \in \alpha$ and $(a\alpha,a) \in \alpha^{-1}$; hence $(a\alpha)\alpha^{-1} = a$. Now $ai_A = a = (a\alpha)\alpha^{-1} = a(\alpha\alpha^{-1})$ for all $a \in A$, and clearly dom $\alpha\alpha^{-1} = $ dom i_A. Therefore $\alpha\alpha^{-1} = i_A$.

9. Let $a \in (X \cap Y)\beta$. Then there exists $z \in (X \cap Y)$ such that $z\beta = a$. Now $z \in X$ and $z \in Y$. Thus $z\beta \in X\beta$ and $z\beta \in Y\beta$; that is, $a = z\beta \in (X\beta \cap Y\beta)$.

A sufficient condition for equality to hold is for the map β to be 1-1.

Section 4

3. Define $\alpha: Z \to Z$ and $\beta: Z \to Z$ by $n\alpha = n + 1$ and $n\beta = 2n$ for all $n \in Z$. Clearly $\alpha, \beta \in M(Z)$. Now $3(\alpha\beta) = (3\alpha)\beta = 4\beta = 8$ and $3(\beta\alpha) = (3\beta)\alpha = 6\alpha = 7$. Thus $\alpha\beta \neq \beta\alpha$.

Section 5

3. $r_2 = r_2 * r_1 = r_1 * r_2 = r_1$.

4. There exist $b,c \in A$ such that $b * a = e$ and $a * c = e$. Now $b = b * e = b * (a * c) = (b * a) * c = e * c = c$.

6. (a) Yes. $a * b = a + b - ab = b + a - ba = b * a$.
   (b) Yes. $(a * b) * c = (a + b - ab) * c$
   $$= a + b - ab + c - (a + b - ab)c$$
   $$= a + b - ab + c - ac - bc + abc$$
   $$= a + b + c - bc - ab - ac + abc$$
   $$= a + b + c - bc - a(b + c - bc)$$
   $$= a * (b + c - bc)$$
   $$= a * (b * c).$$
   (c) Yes. $a * 0 = a + 0 - a \cdot 0 = a$ and $0 * a = 0 + a - 0 \cdot a = a$. Thus 0 is an identity.
   (d) No.

Section 6

2.

$\boxdot$	e	a	b	c
e	e	a	b	c
a	a	e	c	b
b	b	c	e	a
c	c	b	a	e

3. The unique identity is a, $*$ is commutative, $a^{-1} = a$, $b^{-1} = c$, and $c^{-1} = b$.

Section 7

1. (a) Take $q = 0$ and $r = a$.
   (b) Take $q = 1$ and $r = 0$.
3. Let $a \stackrel{\sim}{m} b$ and $x \in Z$. Then there exists $k \in Z$ such that $a - b = km$. Now $a - b + x - x = km$ and $(a + x) - (b + x) = km$. Therefore $(a + x) \stackrel{\sim}{m} (b + x)$. In addition, $(a - b)x = kmx$ and $ax - bx = (kx)m$; thus $ax \stackrel{\sim}{m} bx$.
6. $\bar{1}$ and $\bar{3}$.

CHAPTER 3

Section 1

3. (a), (c), (d)
9. $(a,b) \odot (1,0) = (a,b)$ for all $(a,b) \in G$; therefore $(1,0)$ is a right identity for $\odot$. However, $(1,0)$ is not an identity for $\odot$ since it is not a left identity.
11. Let $A,B \in P(S)$. Then $A,B \subseteq S$ and $(A + B) \subseteq S$, so $(A + B) \in P(S)$. Clearly, associativity and commutativity hold.
$$A + \emptyset = (A \cup \emptyset) \setminus (A \cap \emptyset)$$
$$= A \setminus \emptyset$$
$$= A.$$
Hence $\emptyset$ is the identity. Furthermore, A is its own inverse, since
$$A + A = (A \cup A) \setminus (A \cap A)$$
$$= A \setminus A$$
$$= \emptyset.$$
13. $S \neq \emptyset$; thus there is an element $a \in S$. Now $a * x = a$ has a solution, say e. To show that e is the identity for $*$, let b be an arbitrary element in S. Then $x * a = b$ has a solution, say c; that is, $c * a = b$. Now
$$b * e = (c * a) * e$$
$$= c * (a * e)$$
$$= c * a$$
$$= b.$$
Similarly, $e * b = b$; thus e is the identity. Also, for arbitrary $b \in S$, $b * x = e$ has a solution, say h. Now $h = b^{-1}$, since h is also a left inverse of b. To see this, observe that $x * b = e$ has a solution, say h'. Now
$$h' = h' * e = h' * (b * h)$$
$$= (h' * b) * h$$
$$= e * h = h.$$

Section 2

3. Let $\beta: G_1 \to G_2$ be an onto homomorphism and let $g',h' \in G_2$. Since β is onto, there exist $g,h \in G_1$ so that $g\beta = g'$ and $h\beta = h'$. Now
$$g' \odot h' = (g\beta) \odot (h\beta)$$
$$= (g * h)\beta$$
$$= (h * g)\beta$$
$$= (h\beta) \odot (g\beta)$$
$$= h' \odot g'.$$

5. Suppose that $x\alpha = y\alpha$. Then $\ln x = \ln y$ and $x = y$. Hence α is 1–1. To show that α is onto, let $t \in F$. Since $\ln e^t = t$, e^t is the pre-image of t. For the homomorphism property, let $a,b \in F_+$. Then
$$(ab)\alpha = \ln ab$$
$$= \ln a + \ln b$$
$$= (a\alpha) + (b\alpha).$$

8. The other isomorphism corresponds $\bar{0}$ to 1, $\bar{1}$ to $-i$, $\bar{2}$ to -1, and $\bar{3}$ to i.

9. Suppose that $a\phi = b\phi$. Then $g^{-1} * a * g = g^{-1} * b * g$, which implies $a = b$. Hence ϕ is 1–1. To show that every element in G has a pre-image, let $h \in G$. Then
$$(g * h * g^{-1})\phi = g^{-1} * (g * h * g^{-1}) * g$$
$$= (g^{-1} * g) * h * (g^{-1} * g)$$
$$= h$$
so $g * h * g^{-1}$ is the pre-image of h and ϕ is onto. For the homomorphism property, let $a,b \in G$. Then
$$(a * b)\phi = g^{-1} (a * b) * g$$
$$= (g^{-1} * a) * (b * g)$$
$$= (g^{-1} * a) * (g * g^{-1}) * (b * g)$$
$$= (g^{-1} * a * g) * (g^{-1} * b * g)$$
$$= (a\phi) * (b\phi).$$

Section 3

4. For all $h \in G$ we have
$$h\alpha_{x*y} = h * (x * y)$$
$$= (h * x) * y$$
$$= (h\alpha_x) * y$$
$$= (h\alpha_x)\alpha_y$$
$$= h(\alpha_x\alpha_y).$$
Hence $\alpha_{x*y} = \alpha_x\alpha_y$.

7. Define $\gamma: Z \to S$ by $n\gamma = \beta_n$ for each $n \in Z$. Clearly γ is onto. To verify that γ is 1–1, let $n\gamma = m\gamma$; that is, let $\beta_n = \beta_m$. Now $0\beta_n = 0\beta_m$; thus $0 + n = 0 + m$ and $n = m$. Hence γ is 1–1. Finally, γ possesses the homomorphism property because $m,n \in Z$ imply
$$(m + n)\gamma = \beta_{m+n}$$
$$= \beta_m\beta_n$$
$$= (m\gamma)(n\gamma).$$

Section 4

7. We will utilize 3.4.3.3. The identity e is in H; thus $H \neq \emptyset$. Let $a,b \in H$. Then $a\alpha, b\alpha \in H'$ and $(a\alpha)(b\alpha)^{-1} \in H'$. Since $(a\alpha)(b\alpha)^{-1} = (a\alpha)(b^{-1}\alpha)$ $= (ab^{-1})\alpha \in H'$, we have $ab^{-1} \in H$ and H is a subgroup of G.

8. Suppose, first, that ker $\alpha = \{e\}$. We need to show that α is 1–1. Let $a\alpha = b\alpha$. Then $(a\alpha)(b\alpha)^{-1} = e\alpha$, the identity in G', and $ab^{-1} \in$ ker α. This implies that $ab^{-1} = e$ and $a = b$.

 Conversely, let α be an isomorphism and let $g \in$ ker α. Then $g\alpha = e\alpha$; and since α is 1–1, $g = e$. Hence ker $\alpha = \{e\}$.

Section 5

5. The left cosets of H in G are pairwise disjoint; thus H is the only left coset with the identity and hence the only subgroup.

6. Let $g \in H$. Then $gh \in H$ for each $h \in H$ and $gH = H$. Conversely, let $gH = H$. Then since $e \in H$, $g = ge \in gH = H$.

Section 6

1. $[Z:S] = 6$ and $[E:S] = 3$.

2. $|HK| = 4$ for all possible choices of H and K. Now 4 does not divide 6. Thus HK is not a subgroup of S_3.

8. Let L and R denote, respectively, the sets of left and right cosets of H in G. Define $\beta: L \to R$ by $(xH)\beta = Hx^{-1}$ for all $xH \in L$. Clearly β is onto. We now show that β is 1–1. Let $(xH)\beta = (yH)\beta$. Then Hx^{-1} $= Hy^{-1}$ and $Hx^{-1}y = H$. Hence $x^{-1}y \in H$ by Exercise 6 in Section 3.5 and $xH = yH$ by 3.5.3.3. Now β is 1–1 and onto. Thus there are the same number of left cosets of H in G as there are right cosets of H in G.

Section 7

5. Let G be a cyclic group. Then there exists $x \in G$ such that $G = \langle x \rangle$. Let $a,b \in G$. Then $a = x^m$ and $b = x^n$ for some integers m and n. Now
$$ab = x^m x^n = x^{m+n} = x^{n+m} = x^n x^m = ba.$$

8. Let $x' \in G'$. There exists $x \in G$ such that $x\phi = x'$. Now, for some integer n, we have $g^n = x$ and
$$x' = x\phi = g^n\phi = (g\phi)^n.$$
Therefore each element of G' is an appropriate power of $g\phi$; that is, $G' = \langle g\phi \rangle$.

12. Let G be a group of prime order and let $x \in G$ with $x \neq e$, the identity in G. Now $|x|$ divides $|G|$; thus $|x| = |G|$ and $G = \langle x \rangle$.

14. Let $a,b \in G$. Then $a^2b^2 = (ab)^2$; if and only if $aabb = abab$, if and only if $abb = bab$, if and only if $ab = ba$.

21. Let G be a nonabelian group and let $x \in G$ with $x \neq e$. Then $\langle x \rangle$ is abelian; thus $\langle x \rangle$ is a proper subgroup of G.

CHAPTER 4

Section 1

1. By 3.4.8, $C(G)$ is a group. Let $g \in G$ and $h \in C(G)$. Then $g^{-1}hg = g^{-1}gh = h \in C(G)$, and, by 4.1.2.5, $C(G)$ is normal.

4. Clearly $H\alpha$ is a subgroup of G'. Let $g' \in G'$ and $h' \in H\alpha$. Then there exist $g \in G$ and $h \in H$ such that $g\alpha = g'$ and $h\alpha = h'$. Now
$$(g')^{-1}h'g' = (g\alpha)^{-1}(h\alpha)(g\alpha) = (g^{-1}hg)\alpha \in H\alpha$$
since H is normal and $g^{-1}hg \in H$. Therefore, by 4.1.2.5, $H\alpha$ is normal in G'.

9. We first need to show that HK is a subgroup of G. Let $hk \in HK$. Then for $h' = k^{-1}hk$ we have $h = kh'k^{-1}$ and $hk = (kh'k^{-1})k = kh' \in KH$. Thus $HK \subseteq KH$. Similarly, $KH \subseteq HK$ and equality holds. Now, by Exercise 3 of 3.6, HK is a subgroup of G. To show normality, let $g \in G$ and $hk \in HK$. Then
$$g^{-1}(hk)g = g^{-1}h(gg^{-1})kg = (g^{-1}hg)(g^{-1}kg) \in HK$$
since H and K are both normal.

14. Let g be fixed in G. We wish to show that $H \approx g^{-1}Hg$. Define $\beta: H \to g^{-1}Hg$ by $h\beta = g^{-1}hg$ for all $h \in H$. Clearly β is an onto mapping. Suppose that $x\beta = y\beta$; then $g^{-1}xg = g^{-1}yg$ and $x = y$. Therefore β is 1–1. Furthermore,
$$\begin{aligned}(xy)\beta &= g^{-1}(xy)g \\ &= g^{-1}x(gg^{-1})yg \\ &= (g^{-1}xg)(g^{-1}yg) \\ &= (x\beta)(y\beta).\end{aligned}$$
Hence β is an isomorphism and $H \approx g^{-1}Hg$.

Section 2

4.

	N	aN	a^2N
N	N	aN	a^2N
aN	aN	a^2N	N
a^2N	a^2N	N	aN

6. Let $aH, bH \in G/H$. Then
$$aHbH = abH = baH = bHaH.$$

8. Let $\langle g \rangle = G$ and let aN be arbitrary in G/N. Now $a \in G$; thus there is an integer n such that $g^n = a$. Hence
$$aN = g^nN = (gN)^n \quad \text{and} \quad \langle gN \rangle = G/N.$$

11. Let N be a normal subgroup of G with $[G:N] = p$, where p is a prime. Then G/H is of prime order and has no proper subgroups. Therefore, by 4.2.5, G contains no proper subgroup that contains N.

Section 3

4. $\dfrac{Z_{12}}{\{\bar{0}\}}$, $\dfrac{Z_{12}}{\{\bar{0},\bar{6}\}}$, $\dfrac{Z_{12}}{\{\bar{0},\bar{4},\bar{8}\}}$, $\dfrac{Z_{12}}{\{\bar{0},\bar{3},\bar{6},\bar{9}\}}$,

   $\dfrac{Z_{12}}{\{\bar{0},\bar{2},\bar{4},\bar{6},\bar{8},\overline{10}\}}$, $\dfrac{Z_{12}}{Z_{12}}$.

5. $\dfrac{G}{\{e\}}$, $\dfrac{G}{G}$.

8. Consider $\bar{G} = \{(g,e') \mid g \in G\}$. Clearly this is a subgroup of $G \times G'$. Let $(a,b) \in G \times G'$ and $(g,e') \in \bar{G}$. Now

$$(a,b)^{-1}(g,e')(a,b) = (a^{-1},b^{-1})(g,e')(a,b)$$
$$= (a^{-1}ga, b^{-1}e'b)$$
$$= (g',e') \in \bar{G}.$$

Therefore $\bar{G}$ is normal in $G \times G'$.

Define $\beta: G \to \bar{G}$ by $g\beta = (g,e')$ for all $g \in G$. Clearly β is 1–1 and onto. To show that β possesses the homomorphism property, let $x,y \in G$. Then

$$(xy)\beta = (xy,e') = (x,e')(y,e') = x\beta y\beta.$$

β is an isomorphism; thus $G \approx \bar{G}$, and $\bar{G}$ is a normal subgroup of $G \times G'$.

Section 4

1. We only need show that $N \subseteq HN$, for obviously $g^{-1}ng \in N$ for all $g \in G$ and $n \in N$ implies $g^{-1}ng \in N$ for all $g \in HN$ and $n \in N$.

   Let $a \in N$. Then $a = ea \in HN$; thus $N \subseteq HN$.

7. Clearly $(H \cup N) \subseteq HN$. Now suppose that K is a subgroup of G such that $(H \cup N) \subseteq K$. We need to show that $HN \subseteq K$.

   Let $hn \in HN$. Now $H \subseteq K$ and $N \subseteq K$, so $h \in K$ and $n \in K$. Since K is a subgroup, $hn \in K$ and $HN \subseteq K$.

CHAPTER 5

Section 1

2. By Exercise 11 in 3.1, $\langle P(S), + \rangle$ is an abelian group. Clearly multiplication (i.e., intersection) is associative and commutative. Also, for each $A \in P(S)$, $A \cap S = A = S \cap A$; thus S is the identity. Finally, multiplication distributes over addition, since

$$A \cdot (B + C) = A \cap [(B \cup C) \setminus (B \cap C)]$$
$$= [A \cap (B \cup C)] \setminus [A \cap (B \cap C)]$$
$$= [(A \cap B) \cup (A \cap C)] \setminus [(A \cap B) \cap (A \cap C)]$$
$$= (A \cap B) + (A \cap C)$$
$$= AB + AC.$$

4.   (a), (d).

9.   **ADDITION**

Identity: $a \oplus 1 = 1 \oplus a = 1 + a - 1 = a$; thus one is the additive identity.

Inverse: $a \oplus (2 - a) = (2 - a) \oplus a = 2 - a + a - 1 = 1$; thus $(2 - a)$ is the additive inverse of a.

Commutative: $a \oplus b = a + b - 1 = b + a - 1 = b \oplus a$.

Associative: $(a \oplus b) \oplus c = (a + b - 1) \oplus c$
$$= a + b - 1 + c - 1$$
$$= a + (b + c - 1) - 1$$
$$= a \oplus (b + c - 1)$$
$$= a \oplus (b \oplus c).$$

MULTIPLICATION

Identity: $a \odot 0 = 0 \odot a = a + 0 - a \cdot 0 = a$; thus 0 is the multiplicative identity.

Commutative: $a \odot b = a + b - ab$
$$= b + a - ba$$
$$= b \odot a.$$

Associative: $(a \odot b) \odot c = (a + b - ab) \odot c$
$$= (a + b - ab) + c - (a + b - ab)c$$
$$= a + (b + c - bc) - a(b + c - bc)$$
$$= a \odot (b + c - bc)$$
$$= a \odot (b \odot c).$$

Distributive property of multiplication over addition:
$$a \odot (b \oplus c) = a \odot (b + c - 1)$$
$$= a + (b + c - 1) - a(b + c - 1)$$
$$= (a + b - ab) + (a + c - ac) - 1$$
$$= (a + b - ab) \oplus (a + c - ac)$$
$$= (a \odot b) \oplus (a \odot c).$$

Section 2

3. $r0 = r(0 + 0)$
$$= r0 + r0.$$
Therefore $r0$ is idempotent and $r0 = 0$.

5. $(-e)(-e) = e \cdot e = e$. This result shows that, in a ring with identity, at least two elements (if $-e \neq e$) are their own multiplicative inverses.

8.   (1) $a + a = (a + a)(a + a)$
$$= a^2 + 2a^2 + a^2$$
$$= a + 2a + a$$
$$= (a + a) + (a + a).$$
Hence $a + a = 0$.

(2) $a + b = (a + b)^2 = a^2 + ab + ba + b^2$
$$= a + ab + ba + b.$$

This result implies $ab + ba = 0$. By property (1) $ab + ab = 0$; thus $ab = ba$.

Section 3

3. $(3 \cdot 5)\alpha = 15\alpha = 30$ and $(3\alpha) \cdot (5\alpha) = 6 \cdot 10 = 60$. Therefore $(3 \cdot 5)\alpha \neq (3\alpha)(5\alpha)$, and α does not preserve products.
    Another reason is that α is not onto.

5. Let $a \in R_1$ such that a^{-1} exists. By 5.3.3, $e\alpha$ is the identity for R_2. Now $e\alpha = (aa^{-1})\alpha = (a\alpha)(a^{-1}\alpha)$ and $(a^{-1}\alpha)(a\alpha) = (a^{-1}a)\alpha = e\alpha$. Hence $a\alpha$ has an inverse.

9. Define $\alpha: R \to Z$ (R is the ring of Example 7 in 5.1) by $0\alpha = \bar{0}$, $a\alpha = \bar{0}$, $b\alpha = \bar{1}$, and $c\alpha = \bar{1}$. Also, $\beta: R \to Z$ (defined by $0\beta = \bar{0}$, $a\beta = \bar{1}$, $b\beta = \bar{0}$, and $c\beta = \bar{1}$) and $\alpha: R \to Z$ (defined by $0\gamma = \bar{0}$, $a\gamma = \bar{1}$, $b\gamma = \bar{1}$, and $c\gamma = \bar{0}$) are homomorphisms.

10. (a) Let $\alpha, \beta \in \text{Aut}(R)$. Then $\alpha\beta$ is 1-1 and onto by Exercise 4 of 2.3. In addition, $\alpha\beta$ preserves addition by Exercise 11 of 3.2. We need to show that $\alpha\beta$ preserves multiplication. Let $x, y \in R$. Then
$(xy)\alpha\beta = ((xy)\alpha)\beta = ((x\alpha)(y\alpha))\beta = ((x\alpha)\beta)((y\alpha)\beta) = (x(\alpha\beta))(y(\alpha\beta))$.
Now $\alpha\beta \in \text{Aut}(R)$ and composition is an operation on R. Clearly $i_R \in \text{Aut}(R)$ and $a^{-1} \in \text{Aut}(R)$ for each $\alpha \in \text{Aut}(R)$. Finally, composition of maps is associative; thus $\text{Aut}(R)$ is a group with respect to composition.
(b) $K \neq \emptyset$ since $i_R \in K$. Let $\alpha, \beta \in K$. Then $a\alpha = a$ and $a\beta^{-1} = a$ for all $a \in S$. Finally, $a(\alpha\beta^{-1}) = (a\alpha)\beta^{-1} = a\beta^{-1} = a$ for all $a \in S$; thus $\alpha\beta^{-1} \in K$ and K is a subgroup of $\text{Aut}(R)$.

Section 4

2. $\alpha_{\bar{0}} = \{(\bar{0},\bar{0}), (\bar{1},\bar{0}), (\bar{2},\bar{0})\}$, $\alpha_{\bar{1}} = \{(\bar{0},\bar{0}), (\bar{1},\bar{1}), (\bar{2},\bar{2})\}$, and $\alpha_{\bar{2}} = \{(\bar{0},\bar{0}), (\bar{1},\bar{2}), (\bar{2},\bar{1})\}$.

3. To show that β is 1-1.

Section 5

1. Let S be a subring of Z such that $1 \in S$. Then
$$n \cdot 1 = \overbrace{(1 + 1 + 1 + \cdots + 1)}^{n \text{ addends}} \in S$$
for all $n \in Z$; therefore $Z \subseteq S$ and $S = Z$.

7. $S_2 = \{\ldots, -4, -2, 0, 2, 4, \ldots\}$ and $S_3 = \{\ldots, -6, -3, 0, 3, 6, \ldots\}$ are subrings of Z. $S_2 \cup S_3$ is not a subring of Z.

10. Let S be a subring of Z. For each $n \in S$, every multiple of n is back in S since S is a subring—that is, $mn \in S$ for all $m \in Z$ and $n \in S$. Therefore S is an ideal of Z.

12. I is an ideal; thus for all $r \in R$, $r \cdot e = r \in I$ and $I = R$.

14. Let $x,y \in aR$. Then there exist $r_1, r_2 \in R$ such that $x = ar_1$ and $y = ar_2$. Now
$$x - y = a(r_1 - r_2) \in aR$$
and
$$xy = (ar_1)(ar_2) = a(r_1 ar_2) \in aR.$$
Therefore aR is a subring of R. For the superclosure property, let $r \in R$. Then
$$rx = xr = (ar_1)r = a(r_1 r) \in aR$$
and aR is an ideal.

Section 6

2. The only quotient rings of R are R/R and $R/\{0\}$. Therefore, by 5.6.4, R/R and $R/\{0\}$ are the only homomorphic images of R up to isomorphism.

8. Let e be the identity in R. Now for all $(a + I) \in R/I$ we have
$$(a + I)(e + I) = ae + I = a + I$$
and
$$(e + I)(a + I) = ea + I = a + I.$$
Hence $e + I$ is the identity for R/I.

Section 7

3. Choose $(x,n) = (-e,1)$; then
$$(e,0)(-e,1) = (e(-e) + 1e + 0(-e), 0 \cdot 1) = (0,0).$$

8. There exists $n \in Z_+$ such that $a^n = 0$. Now for all $b \in R$ we have
$$(ab)^n = a^n b^n = 0b^n = 0.$$
Therefore ab is nilpotent.

Section 8

2. Suppose that a^{-1} exists. Then $a \in I$ and $a^{-1} \in R$; thus $e = aa^{-1} \in I$. Now for all $r \in R$, $r = re \in I$. Hence $I = R$.

4. (a) 1   (b) 1   (c) 2   (d) 5   (e) 1

8. Let I be an ideal of Z. I is a subgroup of Z, and I is cyclic. Therefore there exists $a \in I$ such that $\langle a \rangle = I$ (recall that $\langle a \rangle$ is the subgroup generated by a). Clearly $I = \langle a \rangle = (a)$, and I is a principal ideal.

CHAPTER 6

Section 1

1. Let $a \in R$ such that a^{-1} exists. Suppose that there exists $b \in R$ such that $ab = 0$. Then $b = eb = (a^{-1}a)b = a^{-1}(ab) = a^{-1}0 = 0$. Therefore a is not a zero divisor.

5. Since $a \neq 0$, a^{-1} exists. Thus $x = a^{-1}b$ is a solution. To show uniqueness, let x_1 and x_2 satisfy $ax = b$. Then $ax_1 = ax_2$, and, by cancellation, $x_1 = x_2$.

9. By Exercise 7, a field T contains only the ideals $\{0\}$ and T. $(0) = \{0\}$ and $(e) = T$; thus every ideal of T is a principal ideal.

12. The only finite integral domain that is not a field is the integral domain with one element.

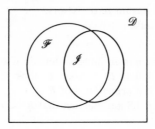

13. Since a is a zero divisor, there exists $r \in R$ such that $ar = 0$ and $r \neq 0$. Now
$$(ab)r = (ba)r = b(ar) = b0 = 0.$$
Therefore ab is a zero divisor.

Section 2

2. (1) → (2) u is an associate of e; thus $u \mid e$ and there exists $d \in D$ such that $ud = e$. Now $du = e$; therefore $d = u^{-1}$.
    (2) → (3) u^{-1} exists; thus $a = ae = a(u^{-1}u) = (au^{-1})u$ and $u \mid a$.
    (3) → (4) Let $a \in D$. Then $u \mid a$; thus there exists $d \in D$ such that $a = ud$ and $a \in (u)$. Hence $(u) = D$.
    (4) → (1) $e \in D$; thus there exists $d \in D$ such that $du = e$. Hence u is a unit.

6. Let a and b be associates. Then there exist $x, y \in D$ such that $a = bx$ and $b = ay$. Now $be = b = ay = (bx)y = b(xy)$, and, by cancellation, we have $xy = e$. Thus x and y are units. Conversely, let u be a unit so that $au = b$. Then $a = bu^{-1}$ and we have $a \mid b$ and $b \mid a$. Thus a and b are associates.

11. $(3,4), (-3,-4), (-4,3), (4,-3)$

12. Every element a different from 0 has an inverse, thus is a unit.

13. When 3 is considered as an integer, it is a prime; and when 3 is considered as a real (or rational) number, it is a unit, not a prime.

16. Let d be a GCD of a and b. There exists $r \in D$ such that $(r) = (a,b)$, since D is a PID. Now $d \mid a$ and $d \mid b$; thus $d \mid r$. Also, since $d \in (a,b)$ and $(a,b) = (r)$, we have $r \mid d$. Hence, by 6.2.6, $(r) = (d)$ and $(d) = (a,b)$.

Conversely, let $(d) = (a,b)$. Clearly $d \mid a$ and $d \mid b$. Suppose that t is such that $t \mid a$ and $t \mid b$. Then $a,b, \in (t)$ and $(a,b) \subseteq (t)$. Hence $(d) \subseteq (t)$, $t \mid d$, and d is a GCD of a and b.

Section 3

1. Let (0) in R be prime. Suppose, by way of contradiction, that R has a zero divisor, say r. Then there exists $t \in R$ with $t \neq 0$ such that $rt = 0$. Now $r \notin (0)$ and $t \notin (0)$, but $rt \in (0)$. This contradicts the fact that (0) is prime; hence R has no zero divisors.
   The converse is trivial.

4. (6)

6. (1) By 6.3.4, $R/\ker \eta$ is an integral domain if and only if $\ker \eta$ is a prime ideal of R. Recall that $R' \approx R/\ker \eta$; thus R' is an integral domain if and only if $\ker \eta$ is a prime ideal of R.
   (2) Follows from 6.3.8 and the fact that $R' \approx R/\ker \eta$.

Section 4

3. When the set A was defined.

8. Let R be the ring of even integers. R has no identity; but, by Exercise 5, R has a field of quotients E. Now
$$E = \{a/b \mid a \in R \quad \text{and} \quad b \in R \setminus \{0\}\}.$$
Note that we can write
$$E = \{2n/2m \mid n \in Z \quad \text{and} \quad m \in Z \setminus \{0\}\}.$$
Define $\beta: Q \to E$ by $(n/m)\beta = 2n/2m$ for all $n \in Z$ and $m \in Z \setminus \{0\}$. Clearly β is 1–1 and onto. To verify the homomorphism properties, let $a/b, c/d \in Q$. Then
$$\left(\frac{a}{b} + \frac{c}{d}\right)\beta = \left(\frac{ad + bc}{bd}\right)\beta = \frac{2(ad + bc)}{2bd} = \frac{(2a)(2d) + (2b)(2c)}{(2b)(2b)}$$
$$= \frac{2a}{2b} + \frac{2c}{2d} = \left(\frac{a}{b}\right)\beta + \left(\frac{c}{d}\right)\beta$$
and
$$\left(\frac{a}{b} \cdot \frac{c}{d}\right)\beta = \frac{2ac}{2bd} = \frac{(2a)(2c)}{(2b)(2d)} = \frac{2a}{2b}\frac{2c}{2d} = \left(\frac{a}{b}\right)\beta \cdot \left(\frac{c}{d}\right)\beta.$$
Therefore β is an isomorphism and $E \approx Q$.

9. By Exercise 8, E and Q are isomorphic; however, R (the ring of even integers) is not isomorphic to Z.

12. Z_n contains zero divisors for any composite n. By Exercise 11, these zero divisors would have to correspond to zero divisors under any isomorphism. Therefore Z_n cannot be isomorphic to a subring of a field; that is, Z_n cannot be imbedded in a field.

Section 5

3. (1) Let $a > e$. Since $e > 0$, we have $a > 0$ by 6.5.4.3. Now $a > e$ and $a > 0$; thus $a \cdot a > a \cdot e$ $(a^2 > a)$ by 6.5.5.3.

   (2) Let $a > 0$ and $ab > ac$. Then $ab - ac > 0$ and $a(b - c) > 0$. Since $a > 0$, we must have $(b - c) > 0$; thus $b > c$.

6. Clearly $\langle e \rangle$ is an integral domain. Let D_+ be the positive set of D. Take $D'_+ = \langle e \rangle \cap D_+$. We will show that D'_+ is a positive set for $\langle e \rangle$. Since $e \in D_+$, we have $e \in D'_+$ and $D'_+ \neq \emptyset$. Let $a,b \in D'_+$. Then $a = ne$ and $b = me$ for appropriate $n,m \in Z$. Now $a + b = ne + me = (n + m)e$ $\in \langle e \rangle$ and $ab = (ne)(me) = (nm)e \in \langle e \rangle$. Since $a,b \in D'_+$, we have a,b $\in D_+$; thus $(a + b)$, $ab \in D_+$. Hence $(a + b)$, $(ab) \in D'_+$, and property (1) of 6.5.1 holds. To verify property (2) of 6.5.1, let $a \in \langle e \rangle$. Suppose first that $a \in D'_+$. Then $a \in D_+$ and, consequently, $a \neq 0$ and $-a \notin D'_+$. Now suppose that $-a \in D'_+$. Then $-a \in D_+$, $-a \neq 0$, and $a \notin D'_+$. Finally, if $a = 0$, clearly $a \notin D'_+$ and $(-a) \notin D'_+$. Therefore $\langle e \rangle$ is an ordered subdomain of D.

7. Let $a \in D_+$, a positive set of an ordered integral domain D. Then $a > 0$ and $2a > a$; thus a cannot be a largest element for D. Clearly $a = 0$ or $a < 0$ cannot be such an element. Therefore D has no largest element.

10. Let $a/e > b/e$. Then
$$\frac{a}{e} - \frac{b}{e} = \frac{ae - be}{e} = \frac{a - b}{e} > 0$$
and $a - b > 0$ since $e > 0$. Therefore $a > b$.
Suppose that $a > b$. Then $a - b > 0$ and
$$\frac{a - b}{e} = \frac{a}{e} - \frac{b}{e} > 0.$$
Hence $a/e > b/e$.

Section 6

2. $n\bar{1} = \bar{n} = \bar{0}$; thus Char $Z_n = n$.

7. By 6.6.6, Z_p, for some prime p, is imbedded in D. Clearly Char $D < 9$. Now p must divide 9. Therefore $p = 3 =$ Char D.

9. Every subfield of Q contains 1, thus Q itself.

10. Yes—the additive order of $\bar{2}$ is 3.

Section 7

2. Let $\alpha \in Z[2]$. Then $\alpha = a_0 + a_1 2^1 + a_2 2^2 + a_3 2^3 + \cdots$, where a_0, $a_1, a_2, \ldots \in Z$. Clearly $a_i 2^i \in Z$ for appropriate i. Hence $\alpha \in Z$ and $Z = Z[2]$.
   Let $\beta = 1 + 3(\frac{1}{2}) + 4(\frac{1}{2})^2$. $\beta \notin Z$; thus $Z \subset Z[\frac{1}{2}]$.

5. Let $\alpha = 1 + 2x - 3x^2 + 4x^3$ and $\beta = 2 - x + 4x^2 - 4x^3$.
   $\deg (\alpha + \beta) = \deg (3 + x + x^2) = 2 < 3 = \max [\deg (\alpha), \deg (\beta)]$.

6. Let $\alpha = \bar{2}x + \bar{3}x^2$ and $\beta = \bar{2}x^3$. $\deg(\alpha\beta) = \deg(\bar{4}x^4 + \bar{0}x^5) = \deg(\bar{4}x^4) = 4 < \deg(\alpha) + \deg(\beta) = 2 + 3 = 5$.

10. $2x \in A$; thus $A \neq \emptyset$. Let $\alpha, \beta \in A$. Then $a_0 > 0$ and $b_0 > 0$, where a_0 and b_0 are, respectively, the leading coefficients of α and β. Now $a_0 + b_0 > 0$ and $a_0 b_0 > 0$; thus $(\alpha + \beta)$, $\alpha\beta \in A$. Let $\gamma \in Z[x]$. Then $\gamma = c_0 + c_1 x + c_2 x^2 + \cdots$. Now exactly one of $c_0 > 0$, $c_0 < 0$, or $c_0 = 0$ holds. Hence exactly one of $\gamma \in A$, $-\gamma \in A$, or $\gamma = 0$ holds and A is a positive set of $Z[x]$.

SYMBOLS

$+_m$	Addition modulo m
Aut G	Automorphism group of G
$A \times B$	Cartesian product set
Char R	Characteristic of a ring R
$\backslash A$	Complement of A
▲	Completion of proof
→	Conditional
$\tilde{m}$	Congruence modulo m
∧	Conjunction
gH, Hg	Cosets (left and right)
$\langle g \rangle$	Cyclic subgroup generated by g
$\deg(f)$	Degree of polynomial f
∨	Disjunction

|	Divides; or Having the property that
dom α	Domain of the mapping α
$\in$	Element of
$\emptyset$	Empty set
$\leftrightarrow$	Equivalence
$\bar{a}$	Equivalence class determined by a
G/H	Factor group
GCD	Greatest common divisor
$\langle G, * \rangle$	Group
(S)	Ideal generated by S
e	Identity, multiplicative
i_A	Identity map on A
$[G;H]$	Index of H in G
Z, Z_0, Z_+	Integers; nonzero integers; positive integers
$\cap$	Intersection
$-a$	Inverse, additive
a^{-1}	Inverse, multiplicative
$\approx$	Isomorphic
ker α	Kernel of a homomorphism
$\alpha: A \to B$	Mapping from A into B
$\cdot_m$	Multiplication modulo m
$\backslash$	Negation
$\lvert a \rvert$	Order of a
$P(A)$	Power set of A
PID	Principal ideal domain
PIR	Principal ideal ring
$\subset$	Proper subset of
R/I	Quotient ring
$S\alpha$	Range of $\alpha: S \to T$
Q, Q_0, Q_+	Rational numbers; nonzero rationals; positive rationals
F, F_0, F_+	Real numbers; nonzero reals; positive reals
$S[x]$	Ring of polynomials in x over S
$\backslash$	Set difference
$M(A)$	Set of mappings on A
$\subseteq$	Subset of
$\mathscr{S}_n$	Symmetric group on n symbols
$\cup$	Union

INDEX